Cambridge
IGCSE®

Mathematics

STUDY AND REVISION GUIDE

Brian Seager, John Jeskins,
Jean Matthews,
Mike Handbury and
Eddie Wilde

HODDER
EDUCATION
AN HACHETTE UK COMPANY

Hachette UK's policy is to use papers that are natural, renewable and recyclable products and made from wood grown in sustainable forests. The logging and manufacturing processes are expected to conform to the environmental regulations of the country of origin.

Orders: please contact Bookpoint Ltd, 130 Milton Park, Abingdon, Oxon OX14 4SB. Telephone: (44) 01235 827720. Fax: (44) 01235 400454. Email: education@bookpoint.co.uk Lines are open from 9 a.m. to 5 p.m., Monday to Saturday, with a 24-hour message answering service. You can also order through our website: www.hoddereducation.com

ISBN: 978 1 4718 5658 7

®IGCSE is the registered trademark of Cambridge International Examinations.

The questions, example answers, marks awarded and/or comments that appear in this book were written by the authors. In an examination, the way marks would be awarded to answers like these may be different.

© Brian Seager, John Jeskins, Jean Matthews, Mike Handbury and Eddie Wilde 2016

First published in 2004

This edition published in 2016 by

Hodder Education,

An Hachette UK Company

Carmelite House

50 Victoria Embankment

London EC4Y 0DZ

www.hoddereducation.com

Impression number 10 9 8 7 6 5 4 3 2 1

Year   2019 2018 2017 2016

Cover photo © senoldo – Fotolia

Illustrations by Integra Software Services

Typeset by Integra Software Services Pvt Ltd, Pondicherry, India

Printed in India

A catalogue record for this title is available from the British Library.

Contents

Introduction

This book is intended to help you prepare for the Cambridge examinations in IGCSE Mathematics, Core and Extended Tiers. The book covers all you need to know for the examination. If you are preparing for the Core, you need only look at the sections shown with a green side band. For the Extended syllabus, you need also to cover the sections shown with a red side band.

The chapters have been arranged in topics: Number; Algebra and graphs; Geometry; Mensuration; Coordinate geometry; Trigonometry; Matrices and transformations; Probability; and Statistics. Each chapter within a topic is introduced in the same way: **Key objectives** outline what will be covered; definitions are supported by **Sample questions** and answers; **Now try this** exercises provide practice; **Sample exam questions** and model answers are provided; and, finally, some **Exam-style questions** are provided for you try. The answers to all the questions are at the back of the book. Throughout there are explanations and **Examiner's tips** to help you make the most of your knowledge and understanding in answering questions in the actual examination.

● How to use this book

There are many ways of using the book. You could use it for reference when doing work from a different book. You could use it by opening a page randomly and checking that you can do the questions. The most satisfactory way, however, will be to use it systematically as part of your planned revision. Whether you choose to work though it in order or jump between the topic areas, you can keep a check on your progress by using the Revision record at the front of the book.

Once you have chosen a topic, see if you can remember how to tackle the various parts. You can either work through the definitions and Sample questions first or try to answer the first question for yourself to test if you remember what it is about. Whichever way you do it, make sure you work through all the exercises and exam questions, as practice reinforces your learning. When you have completed a chapter, don't forget to tick the box in the Revision record. If you know that your understanding is worse in certain topics, don't leave these to the end of your revision programme. Put them in at the start so that you will have time to return to them nearer the end of the revision period.

There are often many ways of tackling a problem. Not all of them will be in this book. If you have learnt a different method that you understand, then keep on using it.

● Examination tips

The examination papers will not tell you the formulae you need to use. You will have to learn and remember them. All those you need to know are in the book. In some cases, you may have learned a formula in a different form. If so, continue to use it.

Make sure you read the instructions carefully, both those on the front of the paper and those in each question. Here are the meanings of some of the words used:

- **Write, write down, state** – little working will be needed and no explanation is required.
- **Calculate, find** – something is to be worked out, with a calculator if appropriate. It is a good idea to show the steps of your working, as this may earn marks, even if your answer is wrong.
- **Solve** – show all the steps in solving the equation.
- **Prove, show** – all the steps needed, including reasons, must be shown in a logical way.
- **Deduce, hence** – use a previous result to help you find the answer.
- **Draw** – draw as accurately and as carefully as you can.
- **Sketch** – need not be accurate but should show the essential features.
- **Explain** – give (a) brief reason(s). The number of features you need to mention can be judged by looking at the marks. One mark means that probably only one reason is required.

Brian Seager, 2016

Cambridge IGCSE Mathematics Study and Revision Guide Second Edition © Brian Seager *et al.*, 2016

1 Numbers and language

Key objectives

- To identify and use natural numbers, integers, prime numbers, square numbers, common factors and common multiples, rational and irrational numbers and real numbers.
- To calculate squares, square roots, cubes and cube roots of numbers.
- To use directed numbers in practical situations.

● Natural numbers and integers

Natural numbers are the counting numbers 1, 2, 3, 4, … and zero. They are positive whole numbers and zero.

Integers are the positive and negative whole numbers and zero. So they are: …, −4, −3, −2, −1, 0, 1, 2, 3, 4, …

● Multiples and factors

The multiples of a number are the numbers in its times table.
The factors of a number are the numbers that divide into it exactly.

Sample questions

1 Find the multiples of 6 up to 30.

2 Find the factors of 36.

Answers

1 6, 12, 18, 24, 30

2 $1 \times 36, 2 \times 18, 3 \times 12, 4 \times 9, 6 \times 6$
   The factors of 36 are 1, 2, 3, 4, 6, 9, 12, 18 and 36.

Now try this

1 Find the first six multiples of the following
   a 4   b 9   c 15

2 Find the factors of the following
   a 24   b 40   c 32

Examiner's tip

It is easier to find factors by looking for pairs of numbers that multiply to give the number.

● Prime numbers and prime factors

Prime numbers are numbers that have only two factors: 1 and the number itself.
Prime factors are factors that are also prime numbers.

Sample questions

1 Find the first five prime numbers.

2 Write 48 as a product of prime factors.

Answers

1 2, 3, 5, 7, 11

2 $48 = 2 \times 2 \times 2 \times 2 \times 3$ or $2^4 \times 3$

Now try this

3 Write the following as a product of prime factors
   a 36   b 140   c 84

Examiner's tip

Keep breaking the numbers into factors until you end up with prime numbers.

● Highest common factor and lowest common multiple

The highest common factor (HCF) is the largest number that will divide into each of the given numbers.

The lowest common multiple (LCM) is the smallest number that can be divided exactly by each of the given numbers.

Sample questions

1 Find the HCF of 120, 36 and 84.

2 Find the LCM of 120, 36 and 84.

Answers

1 $120 = ② \times ② \times 2 \times ③ \times 5.$

$36 = ② \times ② \times 3 \times ③$
$84 = ② \times ② \times ③ \times 7$
$HCF = 2 \times 2 \times 3 = 12.$

Express each as a product of prime factors.

Choose the factors that are common to all three numbers.

2 $120 = ② \times ② \times ② \times 3 \times ⑤.$

$36 = 2 \times 2 \times ③ \times ③$
$84 = 2 \times 2 \times 3 \times ⑦$
$LCM = 2 \times 2 \times 2 \times 3 \times 3 \times 5 \times 7 = 2520.$

Express each as a product of prime factors.

Choose the largest number of primes.

> **Now try this**
> 4 Find the HCF and the LCM of the following
>   a 15 and 24   b 12 and 40   c 90, 54 and 36

● Directed numbers

Negative numbers are used in practical situations such as temperatures.

Sample question

The temperature at mid-day was 8 °C. At midnight it was 12 °C colder.

What was the temperature at midnight?

Answer

Solution $8\,°C - 12\,°C = -4\,°C$

> **Now try this**
> 5 a The balance in Aisha's bank account was –$20. She paid $50 into her account. How much was in her account then?
>   b The car park of a hotel was below ground on floor –2. Vivek went from the car park to his room on floor 4. How many floors did he go up?

● Squares, cubes and roots

To find the **square** of a number, multiply it by itself: $5^2 = 5 \times 5 = 25$. Here are the squares of integers 1 to 15.

Integer	1	2	3	4	5	6	7	8	9	10	11	12	13	14	15
Square number	1	4	9	16	25	36	49	64	81	100	121	144	169	196	225

As 64 is the square of 8, the positive **square root** of 64 is 8. This is written as $\sqrt{64} = 8$. Note that $(-8)^2$ is also 64.

Examiner's tip

You should memorise these square and cube numbers. You will also need them for finding square and cube roots without a calculator.

Cambridge IGCSE Mathematics Study and Revision Guide Second Edition © Brian Seager *et al.*, 2016

To find the **cube** of a number, multiply it by itself and then by itself again: $5^3 = 5 \times 5 \times 5 = 125$. Here are the cubes of the integers 1 to 10.

Integer	1	2	3	4	5	6	7	8	9	10
Cubes number	1	8	27	64	125	216	343	512	729	1000

If you know that $5^3 = 125$, then you also know that the **cube root** of 125 is 5. This is written as $\sqrt[3]{125} = 5$.

Sample questions

Find without a calculator

1 70^2

2 $\sqrt{144}$

3 $\sqrt[3]{512}$

Answers

1 $70^2 = 70 \times 70 = 4900$

2 $\sqrt{144} = 12$ since $12^2 = 144$

3 $\sqrt[3]{512} = 8$, as $8^3 = 512$

> **Now try this**
> 6 Without a calculator, find the following
> a the square of 40
> b $\sqrt{81}$
> c 6^3
> d $\sqrt[3]{343}$

● Rational and irrational numbers

A **rational** number is a number that can be written as an ordinary fraction, $\frac{m}{n}$, where m and n are integers. These are terminating or recurring decimals.

An **irrational** number cannot be written in this form. Decimals that neither terminate nor recur are irrational numbers. For example, π and $\sqrt{6}$ are irrational.

Real numbers are the set of all rational and irrational numbers. Any number on a number line is a real number.

> **Now try this**
> 7 Here are some numbers. Decide which of these is irrational.
> a 126   b $\sqrt{121}$ c $\frac{\pi}{6}$ d $\sqrt[3]{17}$ e $1.3\dot{6}$ f $\left(\sqrt{3}\right)^2$

● Sample exam questions and answers

1 Which of these numbers is irrational?

$\sqrt[3]{27}, \sqrt{48}, 0.12, 2\pi, 0.\dot{3}$

2 Write 30 as a product of prime factors.

1 Irrational: $\sqrt{48}$ (as 48 is not a square number) and 2π (rational $\times$ irrational = irrational).

$\sqrt[3]{27} = 3 = \frac{3}{1}, 0.12 = \frac{12}{100}$ and $0.\dot{3} = \frac{1}{3}$,

so these are all rational.

2 $30 = 2 \times 3 \times 5$

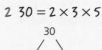

Exam-style questions

1 From the numbers in the range 10 to 20, choose

  a a multiple of 7

  b a factor of 36

  c a prime number

2 Write 264 as a product of prime factors.

3 Find the HCF and LCM of 12 and 16.

4 Find the HCF and LCM of 10, 12 and 20.

5 Find the value of 4 multiplied by the reciprocal of 4. Show clear working to explain your answer.

6 Find the value of $(-5)^2 + 4 \times -3$.

(2) Accuracy

Key objectives

- To be able to make estimates of numbers, quantities and lengths.
- To be able to give approximations to specified numbers of significant figures and decimal places.
- To be able to round off answers to reasonable accuracy in the context of a given problem.

- To be able to give appropriate upper and lower bounds for data given to a specified accuracy.
- To be able to obtain appropriate upper and lower bounds to solutions of simple problems given data to a specified accuracy.

● Accuracy of answers

Sometimes you are asked to round to a given accuracy – either to a given number of decimal places or to a given number of significant figures. Sometimes you need to work to a reasonable degree of accuracy.

● Checking your work

Use the following methods:

- Common sense – is the answer reasonable?
- Inverse operations – work backwards to check.
- Estimates – one significant figure is often easiest.

Sample question

Pat calculated $\dfrac{67.4 \times 504.9}{12.3}$ and got the answer 418 572.2. Use estimates to show her answer is wrong.

Answer

$\dfrac{70 \times 500}{10} = 3500$

Her answer is far too large. (In fact, Pat had multiplied by mistake instead of dividing.)

Now try this

1 Round these to two decimal places (d.p.)
   a 30.972   b 4.1387   c 24.596

2 Round these to three significant figures (s.f.)
   a 5347   b 61.42   c 3049.6

3 Use 1 s.f. estimates to find approximate answers, showing your estimates
   a $792 ÷ 19$ b $\dfrac{5857}{62 \times 20.3}$

● Bounds of measurement

A length measured as 45 cm to the nearest centimetre lies between 44.5 and 45.5 cm. So the bounds are 44.5 and 45.5 cm.

A mass measured as 1.38 kg correct to 3 s.f. has bounds 1.375 and 1.385 kg. As 1.385 rounds up to 1.39 correct to 3 s.f., if we use the bounds to write an inequality for the mass, we have $1.375 \text{ kg} \leq \text{mass} < 1.385 \text{ kg}$.

The lower bound is included in the inequality but the upper bound is not. The bounds are always a half unit down and up in the figure after the degree of accuracy.

Cambridge IGCSE Mathematics Study and Revision Guide Second Edition © Brian Seager *et al.*, 2016

Sample questions

Give the bounds of these measurements. The accuracy of the measurements is given in brackets.

1 80 cm (10 cm)

2 8.6 cm (mm)

3 48 g (g)

4 10.5 s (tenth of a second)

5 5.23 litres (3 s.f.)

Answers

1 75 to 85 cm
2 8.55 to 8.65 cm
3 47.5 to 48.5 g
4 10.45 to 10.55 s
5 5.225 to 5.235 litres

Examiner's tip

A very common error is to give the upper bound in (1) as 84.999 and in (2) as 8.6499, etc.

Now try this

4 Give the bounds of these measurements. The accuracy is given in brackets.
   a 73 g (g)   b 4.3 litre (0.1 litre)   c 7.05 m (3 s.f.)

5 Jack won a race in 25.7 seconds. The time was measured to the nearest tenth of a second. Between what values must the time lie?

● Upper and lower bounds

To find the upper and lower bounds of combined measurements follow these rules:

- Upper bound of a sum, add the two upper bounds.
- Lower bound of a sum, add the two lower bounds.
- Upper bound of a difference, subtract the lower bound from the upper bound.
- Lower bound of a difference, subtract the upper bound from the lower bound.
- Upper bound of a product, multiply the upper bounds.
- Lower bound of a product, multiply the lower bounds.
- Upper bound of a division, divide the upper bound by the lower bound.
- Lower bound of a division, divide the lower bound by the upper bound.

Sample questions

1 A rectangle has sides 3.5 cm and 4.6 cm measured to 2 s.f. Find the minimum and maximum value of

   a the perimeter   **b** the area

2 $A = \dfrac{3b}{c}$, $b = 3.62$, $c = 5.41$ to 2 d.p.

   Find the lower and upper bounds of A.

Answers

1 a Minimum perimeter = $2 \times 3.45 + 2 \times 4.55 = 16$ cm.
     Maximum perimeter = $2 \times 3.55 + 2 \times 4.65 = 16.4$ cm.
  b Minimum area = $3.45 \times 4.55 = 15.6975 = 15.7$ cm^2 to 3 s.f.
     Maximum area = $3.55 \times 4.65 = 16.5075 = 16.5$ cm^2 to 3 s.f.

2 Lower bound of $A = \dfrac{3 \times 3.615}{5.415} = 2.0028 = 2.00$ to 3 s.f.

  Upper bound of $A = \dfrac{3 \times 3.625}{5.405} = 2.0120 = 2.01$ to 3 s.f.

Now try this

6 $P = ab - 2c$, $a = 2.1$, $b = 5.4$ and $c = 3.6$ correct to 2 s.f. Find the lower and upper bounds of the value of P.

7 A rectangle has an area of 14.5 cm² and a length of 4.6 cm, both measured to the nearest 0.1 units. Work out the limits between which the width must lie.

● Sample exam question and answer

When a ball is thrown upwards the maximum height, h, it reaches is given by $h = \dfrac{U^2}{2g}$.

It is given that $U = 4.2$ and $g = 9.8$, both correct to 2 s.f.

Calculate the upper and lower bounds of h.

Upper bound = 0.926. Lower bound = 0.874.

To find the upper bound of h, use the upper bound of U and the lower bound of g. To find the lower bound of h use the opposite.

Upper bound of $h = \dfrac{4.25^2}{2 \times 9.75} = 0.926\,282 = 0.926$ to 3 s.f.

Lower bound of $h = \dfrac{4.15^2}{2 \times 9.85} = 0.874\,238\,5 = 0.874$ to 3 s.f.

Exam-style questions

1 A pane of glass is 3.5 mm thick and measures 1.4 m by 0.9 m. Work out the volume of the glass: (a) in mm³, (b) in m³.

2 The volume of a solid metal cylinder is 600 cm³. Each cm³ of the metal has mass 15 g. Calculate the mass of the metal cylinder in kilograms.

3 A formula used in science is $a = \dfrac{v - u}{t}$. $u = 17.4$, $v = 30.3$ and $t = 2.6$, all measured correct to the nearest 0.1. Find the maximum possible value of a.

4 The population of Kenya is 2.6×10^7, correct to 2 s.f. The area of Kenya is 5.8×10^5 km², correct to 2 s.f. Calculate the lower and upper bounds for the number of people per square kilometre in Kenya.

5 In 1988 a firm produced 1.2 billion litres of fizzy drink (correct to 2 s.f.). The volume of a standard swimming pool is 390 m³ (correct to 2 s.f.). What is the greatest number of these swimming pools that could possibly be filled with this amount of fizzy drink?

Cambridge IGCSE Mathematics Study and Revision Guide Second Edition © Brian Seager *et al.*, 2016

Calculations and order

● Ordering quantities

Here are the symbols you should know:

- $a < b$ means 'a is less than b'.
- $a \leq b$ means 'a is less than or equal to b'.
- $a > b$ means 'a is greater than b'.
- $a \geq b$ means 'a is greater than or equal to b'.

Expressions involving these signs are called inequalities.

Sample questions

Find the integer (whole number) values of x for each of these inequalities

1 $-3 < x \leq -1$

2 $1 \leq x < 4$

Answers

1 If $-3 < x \leq -1$, then $x = -2$ or -1.
   Note that -3 is not included but -1 is

2 If $1 \leq x < 4$, then $x = 1, 2$ or 3.
   Note that 1 is included but 4 is not

● Ordering fractions, percentages and decimals

To put fractions in order of size:

- Change to fractions with the same denominator.
- Order them by the numerators.

Sample question

Put these fractions in order of size, smallest first.
$\frac{1}{2}, \frac{3}{4}, \frac{1}{6}, \frac{2}{3}$

Answer

$\frac{1}{2} = \frac{6}{12}, \frac{3}{4} = \frac{9}{12}, \frac{1}{6} = \frac{2}{12}, \frac{2}{3} = \frac{8}{12}$

Order is $\frac{2}{12}, \frac{6}{12}, \frac{8}{12}, \frac{9}{12}$

$= \frac{1}{6}, \frac{1}{2}, \frac{2}{3}, \frac{3}{4}$

> Fractions changed to denominator 12

This can also be written using inequality symbols: $\frac{1}{6} < \frac{1}{2} < \frac{2}{3} < \frac{3}{4}$.

To order a mixture of fractions, percentages and decimals, it is best to change them all to decimals and put these in order. See the next topic for help in doing this.

● Order of operations

You need to remember the rule that you must multiply and divide before you add and subtract.

When doing complex calculations with a calculator, be safe and introduce brackets. Remember that for every opening bracket there must be a closing bracket.

Examiner's tip

Calculators vary. The position and symbols used on the buttons are different depending on the make and model of calculator. The order in which the buttons have to be pressed can also vary. Make sure you know how **your** calculator works. Do not borrow a different calculator or change calculator just before an exam.

Sample questions

Work out:

1 $2 + 3 \times 4$ **2** $24 - 3 \times 7$

Answers

$1 \quad 2 + 3 \times 4 = 2 + 12 = 14$
$2 \quad 24 - 3 \times 7 = 24 - 21 = 3$

Now try this

8 Work out the following
   a 16×26 b 84×24 c 12×24

9 Work out the following
   a 17×32 b 45×16 c 143×16

10 Work out the following
   a $70 - 4 \times 2$ b $64 + 2 \times 3$ c $3 \times 4 + 5$

11 Work out the following
   a $\dfrac{3.6 \times 5.9}{2.47 - 1.98}$ b $3.2^2 - \sqrt{4.84}$

   c $5.8 \times 1.7 + 3.5 \times 4.4$ d $\dfrac{26.9 - 7.8}{\sqrt{8} + 1.95}$

Sample questions

Work out these

1 $\dfrac{2.3 + 4.5}{5 \times 3.2}$ **2** $6.4 + 7.2 \times 8$

Answers

$1 \quad (2.3 + 4.5) \div (5 \times 3.2) = 0.425$
$2 \quad 6.4 + (7.2 \times 8) = 64$

● Sample exam questions and answers

Work out the following. Give your answers to 2 d.p.

1 4.2^4 **2** $\dfrac{3.9^2 + 0.53}{3.9 \times 0.53}$

$1 \quad 311.17$
$2 \quad 7.61$

Exam-style questions

1 Put these fractions in order of size, smallest first
   $\frac{3}{4}, \frac{7}{16}, \frac{1}{2}, \frac{5}{8}$

2 List the following numbers in order, starting with the smallest
   $66\%, \frac{3}{5}, 0.62, 0.59, 55\%$

3 Work out these
   a $\dfrac{18.6 - 2.75}{3.5 + 1.043}$

   b $\dfrac{1}{4.5 + 6.8}$

   c $3.2\left(5.2 - \dfrac{1}{1.6}\right)$

4 Work out these
   a $\left(\dfrac{4.2^5 - 1.7^4}{1.25^2}\right)$

   b $\sqrt{\left(3^2 + 2.7^3\right)}$

Cambridge IGCSE Mathematics Study and Revision Guide Second Edition © Brian Seager *et al.*, 2016

Integers, fractions, decimals and percentages

4

● Changing between fractions, decimals and percentages

- To change a fraction to a decimal, divide the numerator by the denominator.
- To change a decimal to a percentage, multiply by 100.
- To change a percentage to a decimal, divide by 100.
- To change a percentage to a fraction, write over 100 and cancel.

Sample questions

1 Change $\frac{4}{5}$ to

  a a decimal

  b a percentage

2 Change 35% to

  a a decimal

  b a fraction

Answers

1 a 0.8 $4 \div 5 = 0.8$

  b 80% $0.8 \times 100 = 80$

2 a 0.35 $35 \div 100 = 0.35$

  b $\frac{35}{100} = \frac{7}{20}$ Cancel by 5

Now try this

1 Change these decimals to percentages
  a 0.34   b 0.27   c 0.03   d 0.428

2 Change these fractions to (i) decimals, (ii) percentages
  a $\frac{7}{10}$ b $\frac{3}{5}$ c $\frac{9}{20}$

3 Change these percentages to (i) decimals, (ii) fractions
  a 40%   b 24%   c 5%

● Adding and subtracting fractions

Change the fractions to the same denominator, then add or subtract the numerators.

When mixed numbers are involved, deal with the whole numbers first.

In subtraction, when the first fraction part is smaller than the second, you can change a whole number to a fraction.

Sample question

$3\frac{1}{4} - 1\frac{2}{5}$

Answers

$2 + \frac{1}{4} - \frac{2}{5}$

$= 2 + \frac{5}{20} - \frac{8}{20} = 1 + \frac{20}{20} + \frac{5}{20} - \frac{8}{20} = 1\frac{17}{20}$

Now try this

4 $2\frac{3}{8} + 3\frac{1}{6}$

5 $1\frac{2}{3} + 3\frac{4}{5}$

6 $3\frac{4}{5} - 2\frac{1}{4}$

7 $3\frac{3}{4} - 2\frac{2}{3}$

8 $4\frac{2}{5} - 1\frac{5}{6}$

9 $\frac{3}{8} + \frac{3}{4} - \frac{1}{6}$

Examiner's tip

$1 = \frac{2}{2} = \frac{3}{3}$ etc.

● Multiplying and dividing fractions

When mixed numbers are involved, first change to top-heavy fractions.

To multiply fractions, multiply the numerators and the denominators and then simplify.

To divide fractions, invert the second fraction and multiply.

Sample questions

1 $2\frac{1}{2} \times 1\frac{2}{5}$

2 $3\frac{3}{4} \div \frac{1}{2}$

Answers

1 $\frac{5}{2} \times \frac{7}{5} = \frac{1}{2} \times \frac{7}{1}$ Make fractions top-heavy and

$= \frac{7}{2} = 3\frac{1}{2}$ cancel by 5

2 $\frac{15}{4} \div \frac{1}{2}$ Make fraction top-heavy

$= \frac{15}{4} \times \frac{2}{1}$ Invert the second fraction and multiply

$= \frac{15}{2} \times \frac{1}{1}$ Cancel by 2

$= \frac{15}{2} = 7\frac{1}{2}$

Now try this

10 $2\frac{1}{3} \times 4\frac{1}{2}$

11 $3\frac{1}{3} \times 1\frac{1}{5}$

12 $2\frac{2}{5} \times 4\frac{3}{4}$

13 $2 \div \frac{3}{5}$

14 $3\frac{1}{3} \div \frac{1}{2}$

15 $4\frac{1}{5} \div 2\frac{1}{3}$

● Recurring and terminating decimals

All fractions are equal to either recurring or terminating decimals.

The fractions equal to terminating decimals have denominators whose only prime factors are 2 and/or 5.

Sample questions

Write these fractions as decimals.

1 $\frac{1}{8}$

2 $\frac{1}{6}$

3 $\frac{13}{20}$

4 $\frac{7}{25}$

5 $\frac{3}{7}$

Answers

1 0.125. Terminating, prime factor 2.
2 0.1$\dot{6}$. Recurring, prime factors 2 and 3.
3 0.65. Terminating, prime factors 2 and 5.
4 0.28. Terminating, prime factor 5.
5 0.$\dot{4}$2857$\dot{1}$. Recurring, prime factor 7.

● Changing a recurring decimal to a fraction

The method for doing this is best illustrated by a question.

Sample question

Express $0.4\dot{2}$ as a fraction in its lowest terms.

Answer

Let $r = 0.4\dot{2} = 0.42424242\ldots$

$100r = 42.424242$ Multiply by 100

Subtract: $99r = 42$

$r = \dfrac{42}{99} = \dfrac{14}{33}$

Examiner's tip
- Multiply by 10^n, where n is the number of recurring figures.

Now try this

Express these decimals as fractions in their lowest terms

16   $0.\dot{7}$ 17 $0.6\dot{1}$ 18 $0.41\dot{6}$

● Sample exam question and answer

A railway has a special offer on some fares. This is the advertisement:
What is the cost of the normal fare?

Fares Reduced
$\frac{1}{3}$ off normal fare
You pay only $75.

Fare is reduced by $\frac{1}{3}$.

To find normal fare, divide by $(1 - \frac{1}{3})$.

Normal fare is $\$75 \div \frac{2}{3} = 75 \times \frac{3}{2}$

$= \$112.50$

Exam-style questions

1 Work out, writing the answers as fractions

   a $2\frac{3}{4} + \frac{3}{8}$

   b $3\frac{2}{3} - 2\frac{1}{2}$

2 Work out the following. Give your answers as fractions as simply as possible.

   a $1\frac{3}{5} \times 2\frac{2}{9}$

   b $1\frac{1}{4} + 2\frac{3}{5}$

3 All the prices in Helen's shop are reduced by $\frac{1}{5}$ in a sale.

   a   What is the sale price of a jumper that cost $27.50 before the sale?

   b   The sale price of a dress is $72. What was the price of the dress before the sale?

4 Work out, giving your answers as fractions in their lowest terms

   a $2\frac{3}{8} - 1\frac{1}{2}$

   b $2\frac{1}{4} \div 1\frac{4}{5}$

5   $h = \dfrac{2xy}{x + y}$. Find h when $x = \frac{2}{5}$, $y = \frac{2}{7}$.

6 In an election $\frac{3}{4}$ of the voters chose the Orange party. The number who chose the Orange party was 13 845. How many people voted altogether?

It is illegal to photocopy this page

Key objectives

- To be able to calculate a given percentage of a quantity.
- To be able to express one quantity as a percentage of another.
- To be able to calculate percentage increase or decrease.
- To be able to carry out calculations involving reverse percentages.

● Finding a percentage of an amount

To find a percentage of an amount, change the percentage to a decimal and multiply.

Now try this

1 Work out
   a 15% of $48
   b 37% of 46 m

● Expressing one quantity as a percentage of another

To work out A as a percentage of B, first write A as a fraction of B, then change to a decimal and then a percentage.

To find an increase or decrease as a percentage, use the original value.

Sample question

A coat is reduced in price from $75 to $69. What percentage reduction is this?

Answer

$$\frac{75-69}{75} = \frac{6}{75} = 0.08 = 8\%$$

Now try this

2 Express $14 as a percentage of $40.

3 Express 4 m as a percentage of 32 m.

4 Express 15 litres as a percentage of 25 litres.

5 Haroon's wage increased from $160 to $180 a week. What percentage increase was this?

Examiner's tip

Make sure units are the same.

● Percentage increases or decreases

To find a percentage increase or decrease:

- add or subtract the percentage from 100
- change to a decimal
- multiply.

If an amount increases by 5% each year, multiply by $(1.05)^n$ to find the amount after n years.

If an amount decreases by 5% each year, multiply by $(0.95)^n$ to find the amount after n years.

Sample question

$3500 is increased by 2% per year. What is the amount after 5 years? Give the answer to the nearest cent.

Answer

$100 + 2 = 102\% = 1.02$.   Multiply by 1.02.

After 5 years $= 3500 \times (1.02)^5 = \$3864.2828 = \$3864.28$

Cambridge IGCSE Mathematics Study and Revision Guide Second Edition © Brian Seager *et al.*, 2016

Now try this

6 Increase:
  a $84 by 7%   b $6300 by 2.5%

7 Decrease:
  a $12 by 10%   b $5900 by 3%

8 Find the value of $2000 when it is decreased by 5% each year for 3 years.

9 The insurance value of a house increased by 6% each year. In 2011 it was valued at $152 000. What was its insurance value 4 years later?

● Finding an amount before a percentage increase or decrease

To find the original amount before a percentage increase or decrease:

- add or subtract the percentage to or from 100
- change to a decimal
- divide.

Sample question

The price of a skirt was reduced by 5%. It now costs $27.55. What was the original price?

Answer

Decrease, so new price = 100 − 5 = 95% of original price.

To find original price, divide by 0.95 (95% = 0.95).

Original price = 27.55 ÷ 0.95 = $29.

Now try this

10 A quantity is increased by 3%. It is now 61.8. What was it before the increase?

11 A wage is increased by 6% and is now $8692. What was it before the increase?

12 An amount is decreased by 2.5% and is now $1170. What was the amount before the decrease?

13 A garage decreased the price of all its cars by 7.5%. A car is now priced at $7511. What was the price before the discount?

● Sample exam questions and answers

1 a Jenny used to go to the village hairdresser, where a trim cost $4. The hairdresser left, so she went to a salon in town. A trim there cost $30. What percentage increase is that?

  b At the town salon the prices have gone up several times. The last increase was 6%. She now pays $37.10. How much did she pay before the last increase?

1 a 650%
   Increase = $26. Original was $4. Percentage
   increase = $\frac{26}{4} \times 100$.

  b $35   Previous is wanted, so divide by 1.06.
   New = 106% of previous (1.06).
   Previous = 37.10 ÷ 1.06.

Exam-style questions

1   A calculator was sold for $6.95 plus sales tax. Sales tax was 17.5% at the time. What was the selling price of the calculator including sales tax? Give the answers to the nearest cent.

2   The Candle Theatre has 320 seats. At one performance, 271 seats were occupied. What percentage of the seats were occupied?

3   All clothes in a sale were reduced by 15%.

   a   Pari bought a coat in the sale that was usually priced at $80. What was its price in the sale?

   b   Ali bought a shirt for $32.30 in the sale. What was its price before the sale?

4   In 2013 a store added sales tax at 17.5% to the basic price of furniture. A sofa had a basic price of $650.

   a   What was its selling price inclusive of sales tax?

   b   During a sale the store said 'We pay the sales tax!' What percentage discount on the normal selling price was the store giving?

Cambridge IGCSE Mathematics Study and Revision Guide Second Edition   © Brian Seager *et al.*, 2016

6 Ratio and proportion

● Direct proportion

If quantities vary in direct proportion it means that:

- If you double one quantity, you double the other.
- If you treble one quantity, you treble the other.
- If you halve one quantity, you halve the other, and so on.

Sample question

The total cost of books is directly proportional to the number of books bought. If 30 books cost $240, how much will

1 a 120 books and   **b** 15 books cost?

Answer

1 a 120 = 4 × 30, so 120 books cost 4 × $240 = $960.

  b 15 = $\frac{1}{2}$ × 30, so 15 books cost $\frac{1}{2}$ × $240 = $120.

Alternatively, you can work out the cost of one book: 240 ÷ 30 = $8 first, and then multiply by the number of books. So $C = 8n$ is the equation connecting the number of books (n) and the total cost in pounds (C).

Now try this

1 The time taken to mark students' homework is directly proportional to the number of students. If it takes 2 hours to mark 16 students' homework, how long does it take to mark **a** 4, **b** 32, **c** 40 students' homework?

2 An express train travels 165 metres in 3 seconds. How far would it travel in 8 seconds?

● Ratios – simplest form

To write a ratio in its simplest form, change the parts to the same units and cancel by any common factor. To write as $1:n$, divide the second part by the first.

Sample question

Write these ratios

1 a in their simplest form

  b as $1:n$.

  (i) 4:14   **(ii)** 75 cents : $4   **(iii)** 5 cm : 4 km

Answer

1 a (i) 2:7.   Cancel by 2.
  (ii) 75:400 = 3:16.   Change to cents and cancel by 25.
  (iii) 5:400 000 = 1:80 000.   Change to centimetres and cancel by 5.
  b (i) 1:3.5.   Divide by 2.
  (ii) 1:5.33.   Divide by 3.
  (iii) 1:80 000.

Now try this

3 Write these ratios
  a in their simplest form
  b as $1:n$.
    (i) 4:12
    (ii) 24:9
    (iii) 400 m : 2 km
    (iv) 80 cents : $2

● Using ratios

To share a total amount in a given ratio, first add the parts of the ratio together. Find the multiplier by dividing the total amount by the sum of the ratio parts. Then use the same multiplier for the parts.

Sample question

$150 is shared in the ratio $3:4:5$. How much is each part?

Answer

Total of ratios 12.

Parts are $3 \times 150 \div 12 = \$37.50$.

$4 \times 150 \div 12 = \$50$

$5 \times 150 \div 12 = \$62.50$

Multiplier is $150 \div 12$.

Multiply each part by $150 \div 12$.

Now try this

4 Share $18 in the ratio $4:5$.

5 To make pink paint, red and white are mixed in the ratio $2:5$. How much pink paint can be made if 6 litres of red is used?

6 Share $177 in the ratio $1:2:3$.

● Inverse proportion

Sometimes, as one quantity increases, the other decreases. In such cases, you need to divide by the multiplier, rather than multiply. Questions of this type are examples of **inverse proportion**.

Sample question

If three excavators can dig a hole in 8 hours, how long would it take four excavators to dig the hole?

Answer

Clearly, as more excavators are to be used, the digging will take less time.

The multiplier is $\frac{4}{3}$.

$8 \div \frac{4}{3} = 6$

Divide the known time by the multiplier to find the unknown time.

Now try this

7 A journey takes 18 minutes at a constant speed of 32 kilometres per hour. How long would the journey take at a constant speed of 48 kilometres per hour?

8 It takes a team of eight men 6 weeks to paint a bridge. How long would the painting take if there were 12 men?

9 A journey can be completed in 44 minutes at an average speed of 50 kilometres per hour. How long would the same journey take at an average speed of 40 kilometres per hour?

10 A supply of hay is enough to feed 12 horses for 15 days. How long would the same supply feed 20 horses?

● Increase and decrease by a given ratio

A multiplier method is often used when needing to increase or decrease sizes, for instance enlarging a photo by the ratio 3 : 2.

Sample questions

1 Increase 12 cm by the ratio 5 : 2.

2 Decrease $30 by the ratio 3 : 8.

Answers

1 $12 \, cm \times \frac{5}{2} = 30 \, cm$ (As we need the answer to be an increase, the numerator is the larger of 5 and 2.)

2 $\$30 \times \frac{3}{8} = \$11.25$ (As we need the answer to be a decrease, the numerator is the smaller of 3 and 8.)

> **Now try this**
>
> 11 Increase 48 by the ratio
>   a 5:3, b 9:8, c 6:5.
> 12 Decrease 50 by the ratio
>   a 7:10, b 3:4, c 5:8.

● Sample exam questions and answers

A stew for four people contains 250 g of lentils and 100 g of carrots.

1 Meera makes the recipe for eight people. What weight of lentils does she need?

2 Ali makes the recipe for six people. What weight of carrots does he need?

1 500 g	$\frac{250}{4} \times 8$
2 150 g	$\frac{100}{4} \times 6$

Exam-style questions

1 It takes 100 g of flour to make 15 shortbread biscuits.

 a Calculate the weight of flour needed to make 24 shortbread biscuits.

 b How many shortbread biscuits can be made from 1 kg of flour?

2 Eau de parfum is made from three parts flower extract, eight parts water and nine parts alcohol.

 a How much water does a 250 ml bottle of eau de parfum contain?

 b How many litres of eau de parfum can be made with 45 litres of flower extract?

3 To make a map, a scale of 5 cm : 2 km was used.

 a Aberville and Banchester are 7 km apart. How far are they apart on the map?

 b Write the ratio 5 cm : 2 km as simply as possible.

4 a In a survey of 4000 adults, 2800 said they were in favour of carrying organ donor cards. Work out the percentage that said they were in favour.

 b In another survey 300 people said they did carry donor cards. The ratio of men to women carrying donor cards was 7 : 8. Work out the number of men and the number of women carrying donor cards.

5 On the outward leg of a journey, a cyclist travels at an average speed of 12 kilometres per hour for a period of 4 hours. The return journey took 3 hours. What was the average speed for the return journey?

(7) Indices and standard form

Key objectives

- To understand the meaning and rules of indices.
- To use the standard form $A \times 10^n$, where n is a positive or negative integer and $1 \le A < 10$.

● Indices that are integers

Indices are a short way of writing numbers multiplied by themselves.

Examples

$2^4 = 2 \times 2 \times 2 \times 2 = 16$
$2^3 = 2 \times 2 \times 2 = 8$
$2^2 = 2 \times 2 = 4$
$2^1 = 2 = 2$

Examiner's tip

The 4 in 2^4 is called an index or power.

Continuing the pattern

$2^0 = 1$

$2^{-1} = \dfrac{1}{2^1} = \dfrac{1}{2}$

$2^{-2} = \dfrac{1}{2^2} = \dfrac{1}{4}$

Now try this

Find the value of the following

1 3^4 2 3^0 3 3^{-2} 4 10^0 5 6^3 6 2^{-3}

Examiner's tip

Do not forget that $n^0 = 1$ and $n^{-a} = \dfrac{1}{n^a}$ for any non-zero value of n.

● Multiplying and dividing with indices

Multiplying: $n^a \times n^b = n^{a+b}$ Add the powers.

Dividing: $n^a \div n^b$ or $\dfrac{n^a}{n^b} = n^{a-b}$ Subtract the powers.

Powers $(n^a)^b = n^{ab}$ Multiply the powers.

Sample questions

Simplify each of the following, giving your answers with powers

1 $3^2 \times 3^4$ **2** $2y^4 \times 3y^5$ **3** $3^5 \div 3^3$ **4** $(2^3)^4$

Answers

1 $3^2 \times 3^4 = 3^6$. $2 + 4 = 6$.
2 $2y^4 \times 3y^5 = 6y^9$. $4 + 5 = 9$.

3 $3^5 \div 3^3 = 3^2$. $5 - 3 = 2$.
4 $(2^3)^4 = 2^{12}$. $3 \times 4 = 12$.

Examiner's tip

For adding and subtracting, indices are not so helpful. There is no shortcut rule.

For example, $3^2 + 3^3 = 9 + 27 = 36$.

It is illegal to photocopy this page

Cambridge IGCSE Mathematics Study and Revision Guide Second Edition © Brian Seager *et al.*, 2016

Now try this

Simplify the following, using indices

7 $5^2 \times 5^8$

8 $6^5 \div 6^2$

9 $3^4 \times 3^7 \times 3^{-3}$

10 $(2^{-2})^3$

11 $\dfrac{2^6 \times 2^8}{2^5}$

12 $\dfrac{(5^3)^4}{5^7}$

13 Find the value of x in each of the following

a $5^x = 25$

b $2^x = \frac{1}{8}$

c $3^x = 1$

d $6^x \times 6^3 = 36$

● Standard index form

This is used for large or very small numbers. They are written in the form $a \times 10^n$, where n is an integer and $1 \le a < 10$.

   To multiply or divide numbers in standard form without a calculator, use the rules of indices: $(3 \times 10^6) \times (5 \times 10^8) = 15 \times 10^{8+6} = 15 \times 10^{14} = 1.5 \times 10^{15}$.

Example 1
$3\,500\,000 = 3.5 \times 10^6$

Example 2
$0.00042 = 4.2 \times 10^{-4}$

Examiner's tip

Do not forget–to enter standard form on a scientific calculator, use the **EXP** key, not × 10. For example, to enter 3.5×10^6, press 3.5 EXP 6.

Now try this

Write the following as ordinary numbers

14 3.72×10^5

15 4.8×10^{-4}

Write the following in standard form

16 0.000 058 3

17 75 600 000

18 Write these numbers in standard index form

a 46 000

b 0.0484

c 3 600 000

d 5965

19 Write these numbers out in full

a 8×10^6

b 9.4×10^3

c 7.7×10^1

d 3.03×10^{-4}

20 Evaluate $\dfrac{7.5 \times 10^{-4}}{2.5 \times 10^9}$. Give your answer in standard form.

21 Evaluate $(3 \times 10^{-3}) \times (8 \times 10^7)$. Give your answer in standard form.

22 Work out $(4 \times 10^5) \div (8 \times 10^{-6})$. Give your answer in standard index form.

23 Find the approximate value of $\dfrac{(8.42 \times 10^8) \times (2.83 \times 10^{-4})}{6.074 \times 10^5}$ by writing all the numbers correct to 1 s.f.

Work out, first without a calculator then with one

24 $(8 \times 10^5) \times (1.5 \times 10^8)$

25 $(5 \times 10^7) \div (2 \times 10^{-3})$

Examiner's tip

Many calculators display numbers in standard form as if they were powers. You need to interpret your calculator display and write down the number correctly in standard form.

● Indices that are fractions

Using the laws of indices, $n^{\frac{1}{2}} \times n^{\frac{1}{2}} = n^{\frac{1}{2}+\frac{1}{2}} = n^1 = n$ therefore $n^{\frac{1}{2}} = \sqrt{n}$.

Similarly we can show that $n^{\frac{1}{a}} = \sqrt[a]{n}$ and $n^{\frac{a}{b}} = \sqrt[b]{n^a}$ or $(\sqrt[b]{n})^a$.

Example 1

$64^{\frac{1}{3}} = \sqrt[3]{64} = 4$

Example 2

$125^{-\frac{2}{3}} = \dfrac{1}{(\sqrt[3]{125})^2} = \dfrac{1}{5^2} = \dfrac{1}{25}$ or 0.4

> **Examiner's tip**
> The examples show non-calculator methods. Try to obtain the answers using the power keys on your calculator.

> ### Now try this
>
> Write down the value of these
>
> 26 $36^{\frac{1}{2}}$ 27 $1000^{-\frac{1}{3}}$
>
> Use your calculator to find the following
>
> 28 $90^{\frac{2}{3}}$, correct to 1 d.p.   29 $1024^{-\frac{2}{5}}$
>
> 30 Write down the values of the following
>
>   a $16^{\frac{1}{2}} \times 27^{\frac{1}{3}}$ b $3^{-2} \times 9^0$ c $25^{\frac{1}{2}} \div 100^{\frac{1}{2}}$ d $81^{\frac{1}{2}} \times 3^{-2}$
>
> 31 Write these numbers in ascending order: $4^{\frac{1}{2}}, 4^{-2}, 4^0, 4^{\frac{1}{3}}, 4^2$

> **Examiner's tip**
> Know how to use the root and power buttons on your calculator.

● Sample exam questions and answers

The astronomical unit is used to measure large distances between stars. One astronomical unit is 1.496×10^{11} m.

1 What are 100 astronomical units? Give your answer in standard form.

2 Light travels at 2.998×10^8 m s^{-1}. Write this as an ordinary number.

1 1.496×10^{13} m $100 \, AU = 1.496 \times 10^{11} \times 10^2$

2 $299\,800\,000$ m/s $2.998 \times 10^8 = 2.998 \times 100\,000\,000$

Exam-style questions

1 Evaluate $\dfrac{2^8 \times 2}{2^{-2}}$

   a expressing your answer in the form 2^n

   b expressing your answer in standard form correct to 3 s.f.

2 A radar transmitter sends out a beam of radio waves at a frequency of 24 thousand million pulses per second. Write this figure in standard form.

3 a Work out $(3.0 \times 10^4) \times (6.0 \times 10^3)$, writing the answer in standard form.

   b A terawatt is 10^{12} watts. A power station produces 1.2×10^8 watts. Write this in terawatts.

4 Calculate the difference between 5^{-2} and 2^{-5}. Give your answer in standard form.

5 Write the following as whole numbers or fractions.

   a 9^{-2} **b** 9^0 **c** $27^{\frac{1}{3}}$

6 Work out the exact value of $2^{-3} \times 16^{\frac{1}{2}}$.

Cambridge IGCSE Mathematics Study and Revision Guide Second Edition © Brian Seager *et al.*, 2016

Money and finance

Key objectives

- To be able to calculate using money and convert from one currency to another.
- To be able to use given data to solve problems on personal and household finance involving earnings, simple interest and compound interest.

- To be able to extract data from tables and charts.
- To be able to use exponential growth and decay in relation to population and finance.

● Some common money problems

Now try this

1 Three families share the cost of a meal in the ratio $2:3:4$. The total cost of the meal was $72. How much did each family pay?

2 Jenny had to pay 10% tax on $1520 of her income and 22% tax on $1435. How much tax did she pay altogether?

3 One day the exchange rate was $1 = 11.5 rands.
   a Ali changed $250 into rands. How many rands did he receive?
   b Jan changed 800 rands into dollars. How many dollars did he receive?

4 A shopkeeper bought a crate of 120 oranges. He paid $13. He sold 85 of the oranges at a price of $1 for five. He threw away the others. How much profit did he make?

5 A car was priced at $1700. The dealer offered Mia a discount of $\frac{1}{5}$ of the marked price. Calculate

   a this discount and   b the price Mia would need to pay.

● Simple and compound interest

For **simple interest** accounts, the interest earned is paid to another account. Only the amount originally invested earns interest.

For **compound interest**, the interest earned is added to the account and then the total amount in the account earns interest.

Sample questions

Akua invested $4000 in an account earning compound interest at 5% per year.

1 Calculate the amount in her account after 3 years.

2 Calculate how much less interest she would have had if the account had been at 5% simple interest.

Answers

1 A 5% increase means that the multiplier for each year is 1.05.
  After 3 years, the amount in the account = $4000 × (1.05)³ = $4630.50.
2 For simple interest, the interest each year is 5% of
  $4000 = 0.05 × $4000 = $200.
  So, for 3 years the interest earned would be 3 × £200 = $600.
  Compound interest was $4630.50 − $4000 = $630.50. She would have
  $30.50 less with simple interest.

> ### Now try this
>
> 6 a Said invests $320 for 3 years at 6% simple interest. How much interest does he receive?
>   b Aisha invests $320 for 3 years at 6% compound interest. How much interest does she receive?
>
> 7 Vivek invests $600 for 2 years in a simple interest account. He receives $54 interest in total. What was the annual rate of interest on this account?

● Compound interest formula

The compound interest formula gives the total amount, A, in the account when a sum of money, P, is invested at a rate of r% for n years:

$$A = P\left(1 + \frac{r}{100}\right)^n$$

The sum of money invested is sometimes called the principal. You are expected to know this formula.

> ### Now try this
>
> Work these questions twice, using first the formula, then the multiplier method.
>
> 8 An account offered a rate of 4.6% compound interest per year. Bess invests $2500 in this account. How much is in her account after 5 years?
>
> 9 Bo invested money in an account at 6.3% compound interest. After 4 years he has $957.62 in his account. How much did he invest?

● Exponential growth and decay

Compound interest is an example of exponential growth. The multiplier method can be used to solve any of these problems.

● Use a constant multiplier – greater than 1 for growth, less than 1 for decay.

● Use the $\boxed{x^y}$ or $\boxed{\wedge}$ key.

Sample questions

1 A car depreciates by 30% every year. It cost $9000 new. How much is it worth after 5 years?

2 A population of bacteria increases in number by 6% every hour. By what factor has the population grown after 24 hours?

Answers

1 $9000 \times (0.7)^5 = \$1512$. About $1500

2 $(1.06)^{24} = 4.0489$
   The population has grown approximately four times.

Cambridge IGCSE Mathematics Study and Revision Guide Second Edition © Brian Seager *et al.*, 2016

Now try this

10 Work out these

   a $250 \times (1.03)^{10}$ b $(0.89)^{15}$

   c A radioactive substance loses 5% of its mass every hour. What proportion remains after 24 hours?

11 A population of bats is increasing at a rate of 7.6% a year. In 2014 its population was 3000. At this rate of growth, what will be its population in 2020?

12 A car's value depreciates by 13% per year. If it cost $9990 when new in 2014, what will it be worth in 2020?

13 The number of a certain species of bird has been decreasing by $\frac{1}{20}$ every year. If there were 20 000 in 2005, how many were there in 2015?

● Sample exam questions and answers

Mary went from the UK to Australia. The rate of exchange was £1 = $2.83.

1 Mary changed £250 into Australian dollars. How many dollars was that?

2 In Australia she bought a camera for $130. How much was this in £?

1 $707.50 $250 \times 2.83 = 707.5$

2 £45.94 $130 \div 2.83 = 45.94$

Exam-style questions

1 The table shows the prices of different packs of chocolate bars. Find which pack gives the best value for money.

Pack	Size	Price	
Standard	500 g	$1.15	
Family	750 g	$1.59	
Special	1.2 kg	$2.49	

2 Sasha, John and Dario went to a restaurant. They agreed to split the cost in the ratio 4 : 3 : 5.

   a The food cost $58.80. How much did Dario pay?

   b They paid for the drinks in the same ratio. John paid $5.49. How much did Sasha pay?

3 Frankie invested $4000 in a high-interest savings account. The rate of compound interest was 7% per annum. She left the money in the account for 3 years.

   a How much was in Frankie's account at the end of the 3 years?

   b She decides to leave all the money in the account. After how many *more* years will there be more than $8000 in the account?

4 Leah received a bonus of $900 in 2015. This was a 20% increase on her bonus in 2014. How much was her bonus in 2014?

5 A block of ice is melting. Its volume reduces by 15% each hour. It starts at 80 litres. What is its volume after 6 hours?

9 Time

● Measuring time

The time of day can be given using either a.m. and p.m. or the 24-hour clock.

Sample questions

Meera took 2 hours 24 minutes to drive 160 kilometres from Cape Town to Robertson. She left Cape Town at 10.45 a.m.

1 At what time did she arrive in Robertson?   **2** Calculate her average speed.

Answers

1 $10\,h\,45\,min + 2\,h\,24\,min = 12\,h + 69\,min = 13\,h\,9\,min$. Meera arrived at 1:09 p.m.

2 $24\ minutes = \dfrac{24}{60}\ hours = 0.4\ hours$. Average speed $= \dfrac{Distance}{Time} = \dfrac{159}{2.4} = 66.25\,km\,h^{-1}$.

Now try this

1 Write these times in the 24-hour clock
   a 1:50 a.m.   b 2:40 p.m.   c 11:49 a.m.   d 6:30 p.m.

2 Write these times using either a.m. or p.m. notation
   a 0345   b 1456   c 2340   d 1159

3 A train journey lasts 2 hours and 40 minutes. The train arrives at 10:35 a.m. What time did it start its journey?

4 A cyclist rides 0.6 kilometres in 3 minutes. Calculate her average speed in kilometres per hour.

5 Jan starts from home at 11.25 a.m. She walks 7 km at an average speed of $5\,km\,h^{-1}$. At what time does she finish her walk?

● Sample exam question and answer

Pali leaves home at 6:45 am. He cycles 12 kilometres to school and arrives at 7:34 a.m. Calculate Pali's average speed in kilometres per hour.

He takes 49 minutes. His speed is $\dfrac{12}{49}$ km per minute $= \dfrac{12}{49} \times 60\ km\,h^{-1} = 14.7\,km\,h^{-1}$ to 3 s.f.

Exam-style questions

1 A plane flies 5425 km at an average speed of $800\,km\,h^{-1}$ per hour. How long does its journey take? Give your answer to the nearest minute.

2 A car travels 15 km in 14 minutes. What is its average speed in $km\,h^{-1}$?

Cambridge IGCSE Mathematics Study and Revision Guide Second Edition © Brian Seager *et al.*, 2016

Set notation and Venn diagrams

Key objectives

- To use language, notation and Venn diagrams to describe sets and represent relationships between sets.
- To understand definitions of different types of set.

● One set

Sets are a way of classifying groups of objects and the relationships between them.

Sets are either written out using curly brackets { }, with commas between each item or **element** of the set, or they are described after using a symbol and a colon(:).

- $3 \in S$ means that 3 is in set S.
- Yellow $\notin$ S means that yellow is not in set S.
- The number of elements in set S is written n(S).
- The symbol $\varnothing$ is used for a set that is empty.
- The symbol $\mathscr{E}$ is used for the entire or **universal** set of objects being considered.
- S' is the set of everything in $\mathscr{E}$ that is not in set S. It is called the complement of set S.
- Venn diagrams show sets in a diagram instead of using symbols.
- A Venn diagram shows the universal set as a rectangle, with other sets inside as ovals or circles.

In the Venn diagram opposite, the shaded area represents S'.

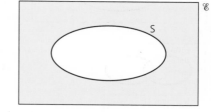

Examples

The set A of integers from 2 to 6 inclusive is

{2, 3, 4, 5, 6} or {x: x is an integer and $1 < x \le 6$}.
In the example above, $4 \in A$, $8 \notin A$.
In the example above, n(A) = 5.

If $\mathscr{E} = \{1, 2, 3, 4, 5, 6, 7, 8, 9, 10\}$ and set A is as above, $A' = \{1, 7, 8, 9, 10\}$.

The Venn diagram opposite shows all the elements of $\mathscr{E}$ and whether they are in A or A'.

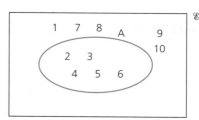

Now try this

1 List the elements of the set {x: $10 \le x \le 20$ and x is an even integer}.

2 Show that $(4, 1) \in \{(x, y): y = 2x - 7\}$.

3 Draw a Venn diagram to illustrate $\mathscr{E}$ = {colours of the rainbow} and S = {red, indigo, violet}.

4 This Venn diagram shows $\mathscr{E}$ and A.
  a List set A'.
  b Write the correct symbol, $\in$ or $\notin$, between each of the following
    (i) 3 A   (ii) 7 A   (iii) 4 A'

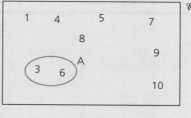

Two or more sets

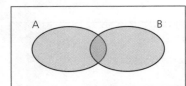

The **union** of sets A and B, written A∪B, is the set of objects belonging to set A or set B or both. It is shown shaded on this Venn diagram.

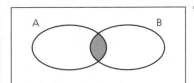

The **intersection** of sets A and B, written A∩B, is the set of objects belonging to both sets A and B.

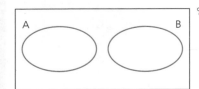

In this Venn diagram, the sets A and B are **disjoint** and the intersection of A and B is the empty set, written A∩B=∅ or A∩B={ }.

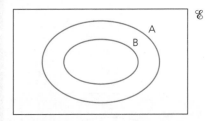

In this Venn diagram, everything in B is in A. B is a **subset** of A. If B is not the same as A, this is called a **proper subset** and written B⊂A. If it is possible for B=A, then it is written B⊆A. B⊄A means B is not a subset of A and has some elements not in A.
   When B≠A, A∪B=A and A∩B=B.

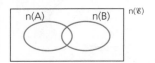

Venn diagrams can show the numbers of elements in each set, instead of the elements themselves. This can be used to solve set problems.

Sample question

In a year group of 72 people, 42 like cricket, 34 like hockey, 18 like both. How many like neither?

Answer

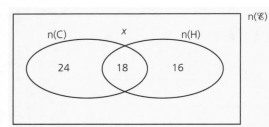

Let x be the number who like neither. Then use n(C) = 42, so n(C∩H) = 24 and so fill in the Venn diagram, starting with n(C∩H') = 18

$x + 18 + 24 + 16 = 72$, so $x = 14$.

Now try this

5 The Venn diagram shows the elements of sets A, B and C. List the elements of these sets
   a A∩C
   b B∩C
   c B∪C
   d A∩C'

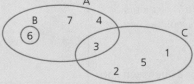

6 Draw a Venn diagram to show sets A, B and 𝒞. Shade the region that represents A∩B'.

7 List the elements of all the possible subsets of A when A={a, b, c}.

Cambridge IGCSE Mathematics Study and Revision Guide Second Edition © Brian Seager *et al.*, 2016

Sample exam questions

1 On this Venn diagram, shade the region that represents $(A \cup B) \cap C'$.

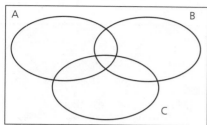

2 Use set language to describe the region that is shaded on this Venn diagram.

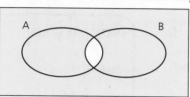

Answers

1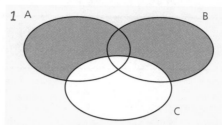

2 *There are two acceptable answers:* $(A \cap B)'$ *or* $A' \cup B'$.

Exam-style questions

1 Three sets A, B and C are such that $A \subset C$, $B \subset C$ and $A \cap B \neq \emptyset$. Draw a Venn diagram to show this information.

2 $\mathscr{E} = \{x: x \text{ is an integer and } 1 \leq x \leq 12\}$, $P = \{x: x \text{ is an even integer}\}$ and $Q = \{3, 6, 9, 12\}$. List the elements of

   a $P \cap Q$ **b** $(P \cup Q)'$

3 In a class of students, 11 play a stringed instrument, 15 play a wind instrument, 6 play both and 10 play neither. Draw a Venn diagram to show this information and find the total number of students in the class.

4 **a** On a copy of this Venn diagram, shade the set $J \cap (K \cup L)$.

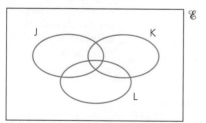

b Express in set notation, as simply as possible, the set shaded in this Venn diagram.

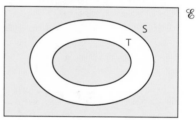

c In a class of 32 students, 12 study chemistry and 18 study French. 9 students in the class study neither of these. Using a Venn diagram, or otherwise, find how many students study chemistry but not French.

Algebraic representation and manipulation

Key objectives

- To use letters to express generalised numbers and express basic arithmetic processes algebraically.
- To substitute numbers for words and letters in formulae.
- To transform simple formulae.
- To construct simple expressions and set up simple equations.
- To manipulate algebraic fractions.
- To factorise and simplify rational expressions.

● Substituting in formulae

Remember that expressions such as ab mean $a \times b$.

Multiplication and division are done before addition and subtraction unless brackets tell you otherwise.

Expressions such as $3r^2$ mean $3 \times r^2$ – which means that you square r and then multiply by 3.

Sample questions

1 If $C = 6b + 3a^2$, find C when $b = -3$ and $a = 5$.

2 If $y = \dfrac{3a - 2b}{c}$, find y when

   a $a = 2$, $b = -5$ and $c = 3$

   b $a = \frac{1}{4}$, $b = \frac{3}{4}$ and $c = 2$

Answers

1 $C = 6 \times -3 + 3 \times 5^2 = -18 + 75 = 57$

2 a $y = \dfrac{3 \times 2 - 2 \times (-5)}{3} = \dfrac{6 + 10}{3} = 5\frac{1}{3}$

   b $y = \dfrac{3 \times \frac{1}{4} - 2 \times \frac{3}{4}}{2} = \dfrac{\frac{3}{4} - 1\frac{1}{2}}{2} = \dfrac{\frac{-3}{4}}{2} = \dfrac{-3}{8}$

Now try this

1 If $A = 3x^2 - 5y^2$, find A when $x = 4$ and $y = -3$.

2 If $y = \dfrac{(a + b)h}{2}$, find y when

   a $a = 2.3$, $b = 5.2$ and $h = 1.9$

   b $a = \frac{1}{2}$, $b = \frac{3}{4}$ and $h = -3$

3 If $s = ut + \frac{1}{2}at^2$, find s when $u = 30$, $a = -10$ and $t = 5$.

● Forming equations

Everyday problems can often be solved by using letters and writing an equation.

Sample question

In a triangle, one angle is twice the size of the smallest angle, and the third angle is $20°$ larger than the smallest angle. Write down an equation and solve it to find the three angles.

Answer

Let the smallest angle be a, one angle be $2a$ and the other $a + 20°$.

So, $a + 2a + a + 20 = 180°$

$$4a = 160°$$

$$a = 40°$$

The angles are $40°$, $80°$ and $60°$.

Examiner's tip

In an exam you must use an equation even if you can work out the answer without doing so. Use it as a check.

Now try this

4 The length of a rectangle is 4 cm greater than its width. The perimeter is 40 cm. Set up an equation and solve it to find the length and width.

5 At a meeting there were 10 more men than women. There were 58 people altogether. Use algebra to work out the number of men and women.

6 A number is squared and 8 added. The result is 89. Set up an equation and solve it to find the number.

Cambridge IGCSE Mathematics Study and Revision Guide Second Edition © Brian Seager *et al.*, 2016

Transforming formulae

To change the subject of a formula, use the same rules as for equations. Do the same thing to both sides to get the new subject on its own on one side of the formula.

Sample question
Make u the subject of $s = \dfrac{(u + v)}{2} \times t$.

Answer

$2s = (u + v) \times t$	Multiply both sides by 2
$2s = ut + vt$	Multiply out the brackets
$2s - vt = ut$	Take vt from both sides
$\dfrac{2s - vt}{t} = u$	Divide both sides by t
$u = \dfrac{2s - vt}{t}$	Rewrite with u on the left

Examiner's tip
Remember always to do the same thing to both sides of the equation.

Now try this

Make the letter in square brackets the subject of these formulae

7   $y = mx + c$ [c]

8   $y = mx + c$ [m]

9   $x = ab - cd$ [b]

10   $x = ab - cd$ [c]

11   $P = 2(i + w)$ [w]

12   $V = \frac{1}{3} i\, wh$ [h]

Powers of the subject

If the required subject is raised to a power, e.g. v^2, first make v^2 the subject and then find the square root of both sides.

For cubes, find the cube root, for power four the fourth root, and so on.

Sample question
Make u the subject of $v^2 = u^2 + 2as$.

Answer

$v^2 - 2as = u^2$	Take $2as$ from both sides
$u^2 = v^2 - 2as$	Change the sides
$u = \sqrt{v^2 - 2as}$	Find the square root of both sides

Subject in twice

If the required subject is in the formula twice:

- Get all the terms involving the subject on one side of the formula and all the other terms on the other side.
- Take the subject out as a common factor.
- Divide both sides by the bracket.

Sample question
Make x the subject of $ax - by = cx + d$.

Answer

$ax = cx + d + by$	Add by to both sides
$ax - cx = d + by$	Take cx from both sides
$x(a - c) = d + by$	Factorise
$x = \dfrac{d + by}{a - c}$	Divide both sides by $(a - c)$

Now try this

Make the letter in square brackets the subject of these formulae. (For example, Question 15 could also be written as 'Find a formula for x in terms of a and y'.)

13 $A = 4\pi r^2$ [r]

14 $y = x^2 + a$ [x]

15 $y = a(x+y)$ [x]

16 $V = \frac{4}{3}\pi r^3$ [r]

17 $y = ax + by$ [y]

18 $ab + cd = ac - bd$ [c]

● Sample exam question and answer

A square has sides of length $x + 2$ cm. The perimeter of the square is 14 cm. Work out the area of the square.

$4(x + 2) = 14$

$4x + 8 = 14$

$4x = 6$

$x = 1\frac{1}{2}$

$Area = 3\frac{1}{2} \times 3\frac{1}{2} = 12\frac{1}{4}$ cm^2.

Exam-style questions

1 The price, P cents, of printing n party invitations is given by $P = 120 + 4n$. Find a formula for n in terms of P.

2 Using $u = 9$, $t = 48$ and $a = -\frac{1}{4}$, work out the value of s from the formula $s = ut + \frac{1}{2}at^2$.

3 a The volume, V, of material inside a tube is given by $V = 25\pi(R^2 - r^2)$. Calculate the value of V when $\pi = 3.1$, $R = 7.3$ cm and $r = 5.9$ cm.

   b Make y the subject of the formula $x = 4y^2 - 3$.

4 Rearrange each of the following to give d in terms of e.

   a $e = 5d + 3$

   b $\frac{3d - 7}{4 + 5d} = e$

5 The formula $f = \frac{uv}{u+v}$ is used in the study of light.

   a Calculate f when $u = 14.9$ and $v = -10.2$. Give your answer to 3 s.f.

   b Express v in terms of u and f.

6 Large radiators cost $45 more than small radiators. Glen buys three small and two large radiators for $415. Write down an equation and solve it to find the cost of a small radiator.

● Multiplying brackets by single terms

When expanding a single bracket, multiply every term within the bracket by the term immediately in front of the bracket.

Now try this

Expand and simplify

19 $a(2a - 3) + 2a(3a + 2)$

20 $4(2x - 3y) - 2(x + y)$

21 $3x(2x + 3) - 4x(x - 3)$

22 $3x + 2x(x - 2)$

Cambridge IGCSE Mathematics Study and Revision Guide Second Edition © Brian Seager *et al.*, 2016

Multiplying two brackets

When expressions of the form $(a+b)(2a-3b)$ are expanded, every term in the first bracket is multiplied by every term in the second bracket.

Sample questions

Multiply out the brackets.

1 $(3a+5)(4a-3)$

2 $(x-3)^2$

Answers

1 $3a(4a-3)+5(4a-3)$
$= 12a^2-9a+20a-15$
$= 12a^2+11a-15$

2 $(x-3)(x-3)$.   Multiply the bracket by itself.
$= x(x-3)-3(x-3)$
$= x^2-3x-3x+9$
$= x^2-6x+9$

Now try this

Multiply out the brackets

23 $(x+5)(x+4)$ 27 $(x+3)(2x-2)$ 31 $(y-2)(y+2)$

24 $(x+7)(x-3)$ 28 $(2x-1)(3x-2)$ 32 $(4x+5)^2$

25 $(x-2)(x+1)$ 29 $(x-3)(5x+4)$

26 $(x-9)(x-3)$ 30 $(a-3b)(a+2b)$

Common factors

Look for every number or letter that is common to every term and write these outside a bracket. Write the terms in the bracket needed to give the original expression when expanded.

Sample questions

Factorise these

1 $3a+6$

2 $2x^2-3xy$

3 $4a-8ac+16a^2b$

Answers

1 $3(a+2)$
2 $x(2x-3y)$
3 $4a(1-2c+4ab)$

Now try this

Factorise these.

33 $5x-25$

34 $3ab-2ac$

35 $2a^2b-6ab^2$

36 $4abc-8ab+2a$

Factorising by grouping

Factorising can be seen as the reverse of expanding brackets.

Sample question

Factorise $12x+16+3xy+4y$.
   There is no factor common to all of these terms so you need to look for pairs of terms that will factorise.

Answers

$(12x+16)+(3xy+4y) = 4(3x+4)+y(3x+4)$

Notice that $(3x+4)$ is a common factor of the two new terms and factorise again:
$= (3x+4)(4+y)$.

You can always check that your answer is correct by expanding it.

Now try this

Factorise these

37 $10x+2xy+3y+15$

38 $2x+3xy+9y+6$

39 $8x+6xy+12+9y$

40 $4b+3ab+20+15a$

41 $7ab+14a+6+3b$

42 $9ab+4+6a+6b$

43 $3x+2xy-15-10y$

44 $2xy+3x-12-8y$

45 $6x-9xy+8-12y$

46 $8-3ab+6b-4a$

● Difference of two squares

You should know that a^2-b^2 factorises to $(a-b)(a+b)$

$(a-b)(a+b)$
$= a^2-ab+ab-b^2$
$= a^2-b^2$

It can be seen from this that when factorising an expression that is a 'difference of two squares' the outcome has the same terms in each bracket with one sign − and the other +.

Examiner's tip
This can be checked by expanding the brackets.

Sample question

Factorise $5x^2-20y^2$.

Answer

$5(x^2-4y^2)$

The terms in the bracket are both squares, factorise.

$= 5(x-2y)(x+2y)$

Now try this

Factorise these

47 x^2-y^2

48 r^2-16s^2

49 p^2-q^2

50 $2a^2-8b^2$

51 v^2-w^2

52 $16m^2-36n^2$

53 x^2-4y^2

54 $5x^2-20y^2$

55 $9a^2-b^2$

56 x^6-y^2

● Factorising quadratics

To factorise a quadratic, first look at the signs in front of the x-term and the number. The examples cover the four different cases. Always multiply out brackets to check.

Sample questions

1 $x^2+5x+6=(x+)(x+)$.
All signs positive.

2 $x^2+x-6=(x+)(x-)$.
Signs different.

3 $3x^2-13x+4=(3x-)(x-)$.
Signs both negative.

4 $2x^2+x-3=(2x)(x)$.
Signs different.

Answers

1 Find two numbers that add to give 5 and multiply to give 6: these are 2 and 3.
$(x+2)(x+3)$

2 Find two numbers that multiply to give −6 and add to give 1: these are 3 and −2.
$(x+3)(x-2)$

3 Find two numbers that multiply to give 12 (3×4) and add to give −13: these are −1 and −12.
$(3x-1)(x-4)$

4 Find two numbers that multiply to give −6 and add to give 1: these are 3 and −2.
$(2x+3)(x-1)$

Now try this

57 x^2+6x+5

58 $x^2-2x-15$

59 $3x^2-5x+2$

60 x^2-6x+8

61 x^2-9

62 x^2-81

63 $3x^2-27$ (Take out the common factor.)

Sample exam questions and answers

1 Expand $(2q-3)(q+5)$.

2 Factorise completely $6p^2-8p$.

1 $2q(q+5)-3(q+5)$
$= 2q^2+10q-3q-15$
$= 2q^2+7q-15$

2 $2p(3p-4)$

Exam-style questions

7 Factorise completely $12p^2q-15pq^2$.

8 Multiply out and simplify these

  a $2(3x+1)-4(x-3)$

  b $(3-x)^2$

9 Expand these

  a $(3x-2)(x+4)$

  b $(2y-1)(y-3)$

10 Factorise these

  a $15a-9-10ab+6b$

  b $3a-6+10b-5ab$

11 Factorise these

  a $50a^2-18b^2$

  b x^4-y^4

12 Factorise these quadratic expressions

  a x^2+x

  b $6x^2-24$

  c x^2-2x-8

  d $2x^2-33x+45$

Manipulating fractions

The rules are the same as for numbers.

Sample questions

Simplify the following

1 $\dfrac{x^2+3x+2}{x^2-1}$

2 $\dfrac{1}{x-1}+\dfrac{1}{x+2}$

Answers

1 $\dfrac{(x+1)(x+2)}{(x+1)(x-1)}$

Factorise the quadratics. Cancel out common factor.

$= \dfrac{x+2}{x-1}$

2 $\dfrac{x+2}{(x-1)(x+2)}+\dfrac{x-1}{(x-1)(x+2)}$

Common denominator is $(x-1)(x+2)$. Multiply top and bottom by necessary factor.

$= \dfrac{x+2+x-1}{(x-1)(x+2)}$

Combine.

$= \dfrac{2x+1}{(x-1)(x+2)}$

Now try this

64 $\dfrac{x^2-x-2}{x^2+2x+1}$

65 $\dfrac{2x^2-8}{2x^2-3x-2}$

66 $\dfrac{1}{x}+\dfrac{2}{x-1}$

67 $\dfrac{x+1}{x-2}-\dfrac{x-2}{x+1}$

● Sample exam question and answer

1 a Factorise $x^2 - 9x + 8$

  b Hence simplify $\dfrac{(x^2 - 9x + 8)}{(x^2 - 1)}$

1 a $(x-1)(x-8)$

  b $\dfrac{(x-1)(x-8)}{(x-1)(x+)} = \dfrac{(x-8)}{(x+1)}$

Exam-style questions

13 Simplify this expression as far as possible: $\dfrac{x^2 + 3x}{x^2 + x - 6}$

14 Simplify

  a $\dfrac{x^2 - 9}{x^2 - x - 6}$

  b $\dfrac{12}{3x + 1} - \dfrac{5}{x + 1}$

Cambridge IGCSE Mathematics Study and Revision Guide Second Edition © Brian Seager *et al.*, 2016

(12) Algebraic indices

● Rules of indices

You need to know the rules of indices. They are:

$$a^m \times a^n = a^{m+n} \qquad \frac{a^m}{a^n} = a^{m-n} \qquad (a^m)^n = a^{m \times n} \qquad a^0 = 1 \qquad \frac{1}{a^n} = a^{-n}$$

$$a^{\frac{1}{2}} = \sqrt{a} \qquad a^{\frac{1}{3}} = \sqrt[3]{a}$$

Remember that $a = a^1$.

Sample questions

1 Write these as whole numbers or fractions

   a $8^{\frac{1}{3}}$ **c** 4^0

   b 2^{-3} **d** $27^{\frac{2}{3}}$

2 Simplify these

   a $3a^2b^2 \times 4ab^3$ **b** $\dfrac{5a^2b \times 4b^3}{12a^3b^2}$ **c** $(2a^2b)^3$

Answers

1 **a** 2   **b** $\dfrac{1}{8}$ **c** 1   **d** 9

2 **a** $12a^3b^5$ **b** $\dfrac{5b^2}{3a}$ **c** $8a^6b^3$

Now try this

Write these as whole numbers or fractions

1 2^3 2 $\left(\frac{1}{3}\right)^{-2}$ 3 $25^{\frac{1}{2}}$ 4 $4^{\frac{3}{2}}$

Simplify these

5 $2abc^2 \times 4ab^2c$ 6 $(4a^2)^3$ 7 $\dfrac{14a^2b^3}{2a^3b}$ 8 $\dfrac{4a \times 3b^2}{2a^2b}$

● Sample exam questions and answers

1 Simplify $\dfrac{2a^4 \times 4a^2}{a^3}$.

2 Find the value of $\left(\frac{49}{4}\right)^{-\frac{3}{2}}$ as a fraction.

1 $8a^3$

2 $\dfrac{8}{343}$

Exam-style questions

1 Simplify these

   a $\dfrac{12p^2q}{3p} \times 2q^2$ **b** $\dfrac{3a^2b \times 4a^2}{2a^3b}$

2 Simplify these

   a $3a^2b \times 4ab^3$ **b** $\dfrac{14a^3b^2}{2ab^3}$

3 **a** Work out the exact value of $2^{-3} \times 16^{\frac{1}{2}}$

   b Simplify as far as possible $\sqrt{p^4q^{-3}} \times \sqrt{\dfrac{q}{p^{-2}}}$

$\overset{\textbf{(13)}}{}$ Equations and inequalities

> **Key objectives**
>
> - To solve simple linear equations in one unknown.
> - To solve simultaneous linear equations in two unknowns.
> - To solve quadratic equations by factorisation, completing the square or by use of the formula.
> - To solve simple linear inequalities.

● Solving equations

Always perform the same operation on each side of the equation.

Sample question

Solve the equation $3x - 4 = 20$.

Answer

First add 4 to both sides giving $3x = 24$. Then divide each side by 3 giving $x = \dfrac{24}{3} = 8$.

> **Now try this**
> 1 $5x + 6 = 31$
> 2 $4p + 4 = 12$
> 3 $4m = 9$
> 4 $7y - 6 = 50$

● Brackets in equations

If the equation has brackets, multiply out the brackets first.

> **Now try this**
> Solve the following equations
>
> 5 $2(m-4) = 10$
> 6 $5(p+6) = 40$
> 7 $7(x-2) = 42$
> 8 $3(4-x) = 21$ (The answer is negative in question 8.)

● Dealing with the 'unknown'

If the 'unknown' is on both sides of the equation, rearrange to collect the numbers on one side and the unknown on the other side.

Sample question

Solve $5x + 4 = 2x + 19$.

Answers

First subtract 4 from each side: $5x = 2x + 15$

Then subtract $2x$ from each side: $3x = 15$, $x = 5$

> **Now try this**
> Solve the following equations
>
> 9 $7x - 4 = 3x + 8$
> 10 $6x - 2 = x + 10$
> 11 $3x + 2 = 8x + 1$
> 12 $5x + 2 = 7x + 10$ (The answer is negative in question 12.)

Cambridge IGCSE Mathematics Study and Revision Guide Second Edition © Brian Seager *et al.*, 2016

● More than one set of brackets

If there is more than one set of brackets, multiply them out first.

Sample question

Solve $5(2x+1)=4(x-2)+10$.

Answer

$$5(2x+1) = 4(x-2)+10$$
$$10x+5 = 4x-8+10$$
$$10x = 4x-8+10-5 = 4x-3$$
$$6x = -3$$
$$x = \frac{-3}{6} = \frac{-1}{2}$$

Now try this

Solve the following equations

13 $6(x+4)=4(x-3)$ (The answer is negative in question 13.)

14 $7(m-2)=5(m+4)+12$

15 $4(p+2)=5(p-5)-9$

16 $3(4x-4)=4(2x+8)$

● Fractions in equations

If the equation has a fraction in it, multiply both sides of the equation by the denominator.

Sample questions

1 $\frac{2}{3}(x+2)=4$

2 $\frac{3}{4}(2-x)=\frac{1}{5}(2x-1)$

Answers

1 Multiply both sides by 3 giving
$$2(x+2) = 12$$
$$2x+4 = 12$$
$$2x = 8$$
$$x = 4$$

2 Multiply both sides by 20
$$15(2-x) = 4(2x-1)$$
$$30-15x = 8x-4$$
$$23x = 34$$
$$x = \frac{34}{23} = 1\frac{11}{23}$$

Now try this

17 $\frac{x}{4}=4$

18 $\frac{3x}{4}=6$

19 $\frac{1}{2}(x+7)=4$

20 $\frac{1}{5}(2x-1)=3$

21 $\frac{1}{2}(x+1)=\frac{1}{3}(2x-1)$

22 $\frac{3}{4}x-2=\frac{x}{3}$

● Sample exam question and answer

Solve the following equations

1 a $\frac{5x+8}{3}=6$

  b $4(x+7)=3(2x-4)$

1 a $\frac{5x+8}{3}=6$
$$5x+8 = 18$$
$$5x = 10$$
$$x = 2$$

b $4(x+7) = 3(2x-4)$
$$4x+28 = 6x-12$$
$$40 = 2x$$
$$x = 20$$

Exam-style questions

1 Solve $3x=x+1$.

2 Solve $3p-4=p+8$.

3 Solve $\frac{3m}{4}=9$.

4 Solve $2(y+3)=5y$.

5 Solve $4(x-1)=2x+3$.

6 Solve $4(x+2)+2(3x-2)=14$.

7 A rectangle has its longer side 2 cm greater than its shorter side. Its perimeter is 36 cm. Let x cm be the length of the shorter side.

   a Write down an equation in x.

   b Solve your equation to find x.

   c Find the area of the rectangle.

● Simultaneous equations

To solve simultaneous equations:

Step 1: Make the coefficients of x or y equal by multiplying one or both of the equations. Do not forget to multiply **every** term.

Step 2: Eliminate the term with equal coefficients:
– if the signs are the same, subtract
– if the signs are different, add.

Step 3: Substitute the value back into one of the original equations to find the value of the other letter.

Step 4: Write down **both** values as your answer.

Step 5: Check your answer by substituting both values back into the equation not used in step 3.

Sample question

Solve the simultaneous equations
$$3x - 5y = 1 \quad (1)$$
$$2x + 3y = 7 \quad (2)$$

Answer

Make y the same by multiplying $(1) \times 3$ and $(2) \times 5$:
$$9x - 15y = 3$$
$$10x + 15y = 35$$
$$19x = 38 \qquad \text{Add}$$
$$x = 2$$
$2 \times 2 + 3y = 7,\ y = 1$ Substitute in (2)
$3 \times 2 - 5 \times 1 = 1$ Check in (1)

● Sample exam question and answer

A gardener finds that two apple trees and five pear trees cost $90 and that three apple trees and six pear trees cost $123. If the cost of one apple tree is x and the cost of one pear tree is y, write down two equations in x and y. Then solve them to find x and y.

$$2x + 5y = 90 \qquad (1)$$
$$3x + 6y = 123 \qquad (2)$$
$$6x + 15y = 270 \qquad (1) \times 3$$
$$6x + 12y = 246 \qquad (2) \times 2$$
$$3y = 24 \qquad \text{Subtract}$$
$$y = 8$$
$$2x + 5 \times 8 = 90 \qquad \text{Substitute } y = 8 \text{ into } (1)$$
$$2x = 50$$
$$x = 25$$
$$x = 25,\ y = 8$$
$$3 \times 25 + 6 \times 8 = 123 \qquad \text{Check in } (2)$$
$$75 + 48 = 123$$

> ### Now try this
>
> 23 Solve these simultaneous equations
>
> a $3x + 2y = 13$
> $4x + y = 14$
>
> b $2x + y = 7$
> $3x - 2y = 28$
>
> c $3x - 2y = 8$
> $4x + 3y = 5$
>
> d $3x - 2y = 7$
> $y = 2 - 4x$
>
> e $x + y = -1$
> $x - y = 6$
>
> f $5x - 3y = 29$
> $3x + 2y = 6$

Exam-style questions

8 Solve these simultaneous equations

$5x + 4y = 13$

$3x + 8y = 5$

9 Solve these simultaneous equations

$4x + 3y = 5$

$2x + y = 1$

10 The equation of the straight line passing through the points (3, 2) and (9, 11) is given by $px + qy = 5$

a Explain why $3p + 2q = 5$

b Write down another equation in terms of p and q.

c Solve the two equations to find the value of p and the value of q.

Cambridge IGCSE Mathematics Study and Revision Guide Second Edition © Brian Seager *et al.*, 2016

● Solving quadratic equations by factorising

If the product of two numbers is 0, then one of the numbers is 0.
   If a quadratic expression equals 0, then one or other of its factors equals 0.

Sample questions

1 Solve $(x-2)(x+3)=0$.
2 Solve $x^2-3x=0$

Answers

1 Either $x-2=0$ or $x+3=0$, so $x=2$ or $x=-3$.
2 $x(x-3)=0$
   Then either $x=0$ or $x-3=0$, so $x=0$ or $x=3$.

Now try this

Solve these quadratic equations by factorising

24 $(x-3)(x-6)=0$ 28 $x^2+3x-10=0$ 32 $x^2+4x=0$

25 $(x+4)(x-1)=0$ 29 $2x^2-13x-7=0$ 33 $2x^2-50=0$

26 $x^2+5x+4=0$ 30 $3x^2-x-10=0$

27 $x^2-7x+12=0$ 31 $6x^2-17x+5=0$

● Solving quadratic equations that do not factorise

There are two ways of doing this algebraically:

1 Completing the square

- If the coefficient of x^2 is not 1, divide through the equation by the coefficient of x^2.
- Make the x^2+bx terms into a complete square and correct the term added by subtracting $\left(\frac{b}{2}\right)^2$.
- Rearrange the equation.
- Take the square root, remembering ±.

Sample question

Solve $2x^2-12x+1=0$.

Answer

$x^2-6x+0.5=0$ Divide by 2

$(x-3)^2-9+0.5=0$ $(x-3)^2=x^2-6x+9$

$(x-3)^2-8.5=0$ $(x-3)^2=8.5$

$x-3=\pm\sqrt{8.5}$

$x=3\pm\sqrt{8.5}$

$=5.92$ or 0.08 to 2 d.p.

2 Using the quadratic formula

- When $ax^2 + bx + c = 0$, $x = \dfrac{-b \pm \sqrt{(b^2 - 4ac)}}{2a}$

- When using this formula, take care with the signs and brackets. Make sure you know how to use your calculator to work it out correctly.

Sample question

Solve $2x^2 - 7x - 3 = 0$.

Answer

$a = 2$, $b = -7$, $c = -3$

$x = \dfrac{7 \pm \sqrt{(-7)^2 - 4 \times 2 \times -3}}{2 \times 2}$

$= \dfrac{7 \pm \sqrt{49 + 24}}{4}$

$= \dfrac{7 \pm \sqrt{73}}{4}$

$= 3.89$ or -0.39 to 2 d.p.

Now try this

Try to solve these by completing the square

34 $x^2 - 4x - 6 = 0$	35 $x^2 + 8x + 2 = 0$	36 $x^2 - 10x + 7 = 0$	37 $x^2 - 3x + 1 = 0$
38 $4x^2 - 8x + 1 = 0$	39 $2x^2 - 4x - 7 = 0$	40 $3x^2 + 12x - 2 = 0$	41 $5x^2 + 30x + 4 = 0$

Try to solve these using the quadratic formula

42 $x^2 - 5x + 3 = 0$	43 $2x^2 + 8x + 1 = 0$	44 $x^2 - 6x + 4 = 0$	45 $3x^2 - 5x - 1 = 0$
46 $4x^2 - 7x + 2 = 0$	47 $2x^2 - 9x - 7 = 0$	48 $3x^2 + 12x + 2 = 0$	49 $5x^2 + 20x + 4 = 0$

50 Try to solve $x^2 - 4x + 7 = 0$ using the quadratic formula or completing the square. What happens? Draw a graph of $y = x^2 - 4x + 7$ and look for $y = 0$.

● Sample exam question and answer

1 a Write $x^2 + 6x + 2$ in the form $(x + a)^2 + b$.

b Hence state the minimum value of y on the curve $y = x^2 + 6x + 2$.

c Solve the equation $x^2 + 6x + 2 = 0$.

1 a $x^2 + 6x + 2 = (x + 3)^2 - 9 + 2 = (x + 3)^2 - 7$

  b The least value of $(x + 3)^2$ is 0, so the least value of y is -7.

  c $x^2 + 6x + 2 = 0$
  $(x + 3)^2 - 7 = 0$
  $(x + 3)^2 = 7$
  $x + 3 = \pm \sqrt{7}$
  $x = -3 \pm \sqrt{7}$
  $x = -0.35$ or -5.65 to 2 d.p.

Cambridge IGCSE Mathematics Study and Revision Guide Second Edition © Brian Seager *et al.*, 2016

Exam-style questions

11 Solve these equations by factorising

   a $x^2 - 6x + 8 = 0$

   b $2x^2 + 3x - 9 = 0$

12 a Factorise completely $5x^2 - 20$.

   b **(i)** Factorise $x^2 - 9x + 8$.

     (ii) Hence solve $x^2 - 9x + 8 = 0$.

13 a Multiply out and simplify the expression $(2x + 7)(3x - 6)$.

   b **(i)** Factorise $x^2 + 6x$.

     (ii) Solve the equation $x^2 + 6x = 0$.

14 Solve the equation $2x^2 - 38x + 45 = 0$.

15 a Write $3x^2 - 12x + 2$ in the form $3(x - a)^2 - b$.

   b Hence solve the equation $3x^2 - 12x + 2 = 0$.

16 The length of a rectangle is y cm, the perimeter is 30 cm and the area is 55 cm².

   a Form an equation in y and show that it can be simplified to $y^2 - 15y + 55 = 0$.

   b Solve the equation $y^2 - 15y + 55 = 0$ to find the length and width of the rectangle. Give your answers correct to 2 d.p. Do not use a trial and improvement method.

17 Bill is making a rectangular chicken pen against a wall. The other three sides will be made from wire netting. Here is the plan.

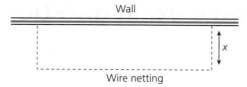

The total length of wire netting is 22 m.

The area inside the pen must be 60 m².

   a Show that $x^2 - 11x + 30 = 0$.

   b Solve the equation.

   c Describe the size of the pen.

● Inequalities

The symbols used are:

- > means greater than.
- < means less than.
- ≥ means greater than or equal to.
- ≤ means less than or equal to.

Inequalities are solved using the same rules as when solving equations **except** when dividing or multiplying by a negative number. Then the inequality sign is reversed.

Sample questions

1 Find the largest integer that satisfies $4x + 10 < 2x + 14$.

2 Solve the inequality $4x < 6x + 2$.

3 Solve $x^2 \leq 16$.

4 Solve $-10 < 2x + 1 \leq 15$.

Answers

1   $4x - 2x < 14 - 10$

      $2x < 4$

       $x < 2$

The largest integer value less than 2 is 1.

2   $-2x < 2$

     $x > -1$

Divided by -2, so reverse the inequality sign.

3   $x \leq 4$ and $x \geq -4$

Turn the inequality round for the negative solution.

The answer $-4 \leq x \leq 4$.

4   $-11 < 2x \leq 14$ Subtract 1 from -10 and 15

   $-5.5 < x \leq 7$

Now try this

51 Solve these inequalities

   a $3x < x + 4$ b $3 - 4x > 6 + x$ c $2(3x - 4) \leq 3(4x - 3)$

52 Find the smallest integer value that satisfies $3(9 - 4x) < 9 + 6x$.

53 Solve $x^2 \geq 25$.

54 Solve $5 < 3 - x < 9$.

● Sample exam question and answer

Solve $2x + 3 < 5x + 12$.

$$2x - 5x < 12 - 3$$
$$-3x < 9$$
$$x > -3$$

Exam-style questions

18 Solve $8x + 5 > 25$.

19 Solve $2x + 17 > 4x + 6$.

20 Solve $x^2 < 36$.

● Graphical solution of a set of inequalities

The solution will be represented by a region of the graph.

Step 1: Draw the boundary lines. Change the inequality sign to an equals sign. Use a continuous line if there is an equals sign in the inequality and a dotted line if there is no equals sign.

Step 2: Substitute the coordinates of the origin, $(0, 0)$, to test each inequality. If the result is true, shade the side of the line opposite the origin. If the result is false, shade the side of the line the same as the origin.

Step 3: The remaining, unshaded area is the region of points satisfying the set of inequalities.

> **Examiner's tip**
>
> Shade the unwanted region unless the question states otherwise.

Sample question

Find the region satisfied by $x \geq 1$, $x + y < 5$ and $y \geq 2x - 6$.

Answer

$x = 1$ is the line through $(1, 0)$ parallel to the y-axis.

The line $x + y = 5$ passes though $(0, 5)$ and $(5, 0)$, and $y = 2x - 6$ passes through $(0, -6)$, $(1, -4)$ and $(2, -2)$.

$x \geq 1 \Rightarrow 0 \geq 1$	False, shade origin side
$x + y < 5 \Rightarrow 0 < 5$	True, shade opposite side
$y \geq 2x - 6 \Rightarrow 0 \geq -6$	True, shade opposite side

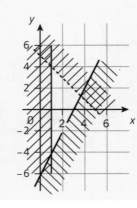

Now try this

55 Find the region satisfied by each set of inequalities

    a $y \leq 4$

       $x + y \geq 4$

       $y \geq 3x - 1$

    b $x \leq 2$

       $2x + 3y \leq 12$

       $5x + 2y \geq 10$

● Sample exam question and answer

1 Write down the three inequalities that satisfy the shaded region.

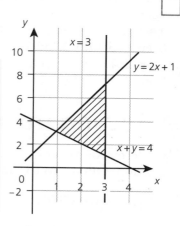

1 Choose a point in the region $x = 2$, $y = 3$.

    $x = 2$

    $x + y = 4$

               $y = 2x + 1$

               $3 \leq 2 \times 2 + 1$

               $2 \leq 3$

            $2 + 3 \geq 4$

               $y \leq 2x + 1$

                $x \leq 3$

             $x + y \geq 4$

Exam-style questions

21 Find the region satisfied by the inequalities

    $x + y \leq 5$

    $y \leq 2x - 1$

    $y \geq 0$

22 Look at the graph and write down the three inequalities that satisfy the shaded region.

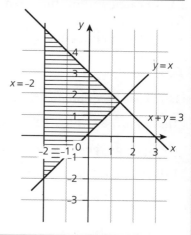

(14) Linear programming

Key objectives

- To represent inequalities graphically.
- To use this representation in the solution of simple linear programming problems.

This deals with problems in which a number of simple conditions have to be satisfied at the same time.

There are three steps in solving a problem by linear programming.

Step 1: Set up the inequalities from the information.

Step 2: Draw the graph.

Step 3: Find the solution that best satisfies the requirements.

● Setting up inequalities

Use a letter for each of the two variables and write an inequality for each restriction. Often two trivial inequalities, $x \geq 0$ and $y \geq 0$, are given.

Sample question

Ahmed buys some doughnuts and some muffins for a party. Doughnuts cost $1.00 each, and muffins cost $1.50 each. He wants at least 12 items. He has $18 to spend. He wants more muffins than doughnuts.

Write down an inequality to represent each of these three conditions.

Answer

Let there be x doughnuts and y muffins.

There must be at least 12 items: $x + y \geq 12$.

Ahmed can spend up to $18 but no more.

$1.0x + 1.50y \leq 18$, so that $2x + 3y \leq 36$.

There must be more muffins than doughnuts: $y > x$.

Now try this

1 A committee must have fewer than 15 members. There must be at least five men and at least five women. There are x men and y women on the committee. Write down three inequalities that x and y must satisfy.

2 A theatre holds 600 people. Adult tickets cost $8 and children's tickets cost $5. At a performance at least $3200 must be taken to make a profit. There must be more adults than children. If x children and y adults are at a performance that makes a profit, write down three inequalities that x and y must satisfy.

● Graphing the inequalities

For each inequality draw the line of the associated equation. Use a continuous line if the line is included in the required region. Use a dotted line if the line is not included in the required region. Shade the unwanted region.

Cambridge IGCSE Mathematics Study and Revision Guide Second Edition © Brian Seager *et al.*, 2016

Sample question

Draw axes taking values of both x and y from 0 to 20 and show by shading the region satisfying the inequalities from the previous sample question: $x + y \geq 12$, $2x + 3y \leq 36$ and $y > x$.

Answer

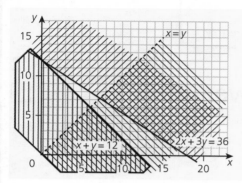

Use a continuous line for $x + y = 12$ and $2x + 3y = 36$.

Use a dotted line for $y = x$.

The required region is the unshaded polygon.

Now try this

For each question, draw and label axes taking values of both x and y from 0 to 12 and show by shading the region satisfied by these inequalities.

3 $y \geq 3$, $4x + 3y \leq 36$, $2x > y$

4 $y \geq 0$, $x < 8$, $y \leq x$, $2x + 3y \geq 18$

5 $4x + 5y \leq 40$, $y \leq 3x$, $y \geq 3$

● Finding the solution

Step 1: Identify the possible solutions in the required region.

Step 2: Choose the one that best satisfies the requirements.

Sample question

In the previous example, find the greatest number of items that can be bought and the cost.

Mark the integer points inside the required region.

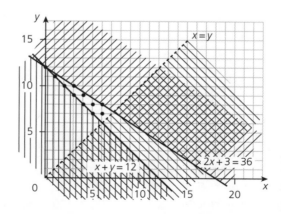

Answer

The greatest number of items is six doughnuts and eight muffins. These cost $18.

Examiner's tip

You do not include (6, 6) or (7, 7) as they are on the dotted boundary.

Now try this

6 a On the graph for question 3, mark all the integer values of (x, y) in the region.
   b (i) Find which of these points gives the greatest value of $x + 2y$.
     (ii) State the value.

7 a Using the graph for question 4, find the integer value of (x, y) in the region that gives the least value of $2x + 3y$.
   b Write down this value.

8 Using the graph for question 5, list the integer values of (x, y) in the region where $x = y$.

● Sample exam question and answer

There are x girls and y boys in a school choir.

1 a (i) The number of girls in the choir is more than 1.5 times the number of boys. Show that $y < \frac{2x}{3}$.

   (ii) There are more than 12 girls in the choir. There are more than five boys in the choir. The maximum number of children in the choir is 35. Write down three more inequalities.

   b (i) Draw axes for $0 \leq x \leq 40$ and for $0 \leq y \leq 40$.
     (ii) Draw four lines on your graph to represent the inequalities in part (a). Shade the unwanted parts of the grid.

   c The school buys a uniform for each choir member. A girl's uniform costs $25. A boy's uniform costs $20. Find the maximum possible cost for the choir uniforms. Mark clearly the point, P, on your graph that you use to calculate the cost.

1 a $x > 1.5y = \frac{3}{2}y$, so $y < \frac{2x}{3}$.
  b $x > 12, y > 5, x + y \leq 35$.

2

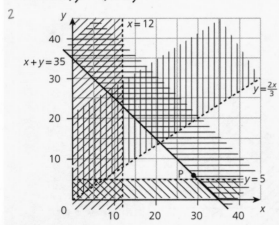

3 $(29, 6)$ marked on graph. Cost $= 29 \times 25 + 6 \times 20 = 725 + 120 = \$845$.

Cambridge IGCSE Mathematics Study and Revision Guide Second Edition © Brian Seager *et al.*, 2016

Exam-style questions

1 A factory produces two types of garden ornaments, type A and type B. Type A takes 1 hour of machine time and 3 hours of workers' time. Type B takes 2 hours of machine time and 1 hour of workers' time. In a day there are 28 hours of machine time and 24 hours of workers' time available.

a If x of type A and y of type B are produced in a day, write down two inequalities satisfied by x and y.

b Draw these inequalities on a graph, shading the unwanted regions.

c The profit made on each type A is $20 and on each type B is $10.

(i) Find the greatest profit that can be made.

(ii) How many of each type are produced to make this profit?

2 A school library is given $1000 for new books. They buy x paperback and y hardback books. There must be at least 50 books and there must be more hardback than paperback books. Paperback books cost $10 and hardback books cost $25.

a (i) Explain why $2x + 5y \leq 200$.

(ii) Write down two more inequalities involving x and y, other than $x \geq 0$, $y > 0$.

b Represent these inequalities on a graph, shading the unwanted regions.

c The school orders the maximum number of hardback books, subject to these conditions. Find this number from your graph.

3 A photographer charges $10 for an individual photo and $30 for a set of six. She always sells more individual photos than sets in a week. She sells x individual photos and y sets of photos in a week. She needs to earn at least $300 a week.

a Write down two inequalities involving x and y, other than $x \geq 0$, $y \geq 0$.

b Represent these inequalities on a graph, shading the unwanted regions.

c To produce an individual photo takes 1 hour and to produce a set takes 2 hours.

(i) What is the least time she can work and earn at least $300?

(ii) How many individual photos and how many sets does she produce in this time?

15 Sequences

Term-to-term rules

Every number in a sequence is called a term of the sequence. There is often a simple rule for going from one term to the next. Use the term-to-term rule to find the formula for a sequence.

Sample question

Here are the first four terms of a sequence: 3, 7, 11, 15.

1 a Find the term-to-term rule.
  b Find the 20th term.
  c Find a formula for the nth term of the sequence.

Answer

1 a Add 4.
  b To get the 20th term, add four
     19 times: $3 + (19 \times 4) = 3 + 76 = 79$.
  c The formula is $3 + (n-1) \times 4$
     $= 3 + 4n - 4 = 4n - 1$.

Now try this

1 Find the first four terms for the sequence given by $3n+3$.

2 Find the term-to-term rule for the sequence 2, 7, 12, 17, ...

3 Find the 15th term and the formula for the nth term for the sequence 5, 11, 17, 23, ...

4 Find the formula for the nth term for the sequence 40, 36, 32, ...

Common sequences

You should know and recognise common sequences.

Examples

The powers of 2: 1, 2, 4, 8, 16, 32, ... 2^n
The powers of 10: 1, 10, 100, 1000, ... 10^n
The square numbers: 1, 4, 9, 16, ... n^2

The triangle numbers: 1, 3, 6, 10 ... $\dfrac{n(n+1)}{2}$

Position-to-term rules

Every number in a sequence can be expressed in terms of its position.
  Use the position-to-term rule to find the formula for the nth term of a sequence:

- look at the difference between successive terms
- multiply this difference by n
- add or subtract a constant to give the required results.

Cambridge IGCSE Mathematics Study and Revision Guide Second Edition © Brian Seager *et al.*, 2016

Sample question

Find the rule for this sequence: 7, 11, 15, 19, … .

Answer

It is sometimes useful to write the sequence out in columns

Position	Term	Difference between terms
1	7	
		4
2	11	
		4
3	15	

This shows that you multiply by 4. The term $4n$ will be in the formula.

Look at position 1. The term is 7. $4 \times n = 4 \times 1 = 4$. You need to add 3 to give the required result of 7.

Check for position 2: $4 \times 2 + 3 = 11$

So the nth term $= 4n + 3$.

Now try this

Find a formula for the nth term of each of these sequences

5 3, 5, 7, 9, …

6 10, 20, 30, 40, …

7 −4, 1, 6, 11, …

8 32, 29, 26, 23, …

● Other sequences

Some sequences do not have a common difference between terms. For example, if each term is the previous term multiplied by a constant, the sequence is a geometric one.

The rules for other sequences can often be found on inspection.

Sample question

Find a formula for the nth term of the sequence 4, 20, 100, 500, … .

Answer

Each term is $5 \times$ the previous one.

The 2nd term $= 4 \times 5$.

The 3rd term $= 4 \times 5 \times 5 = 4 \times 5^2$.

The 4th term $= 4 \times 5 \times 5 \times 5 = 4 \times 5^3$.

$U_n = 4 \times 5^{n-1}$ *The nth term is also written 'U_n'.*

Sample question

Find a formula for the nth term of the sequence 2×3, 3×4, 4×5, … .

Answer

$U_1 = 2 \times 3 = (1 + 1)(1 + 2)$

$U_2 = 3 \times 4 = (2 + 1)(2 + 2)$

$U_n = (n + 1)(n + 2)$

Now try this

9 Find the formula for the nth term of the following sequences

a 2, 8, 32, 128, …

b 1, 5, 25, 125, …

c 3×5, 4×6, 5×7, …

● Sample exam question and answer

1 a Write down the 10th term for the sequence 3, 7, 11, 15, … .

  b Write down an expression for the nth term.

  c Show that 137 cannot be a term in this sequence.

1 a 39

    10th term $= 3 + 9 \times 4$

  b $4n - 1$

    Either: the difference between terms is 4, so the expression will start $4n$. If $n = 1$, then $4n = 4$, therefore subtract 1 to get 3. Therefore the expression is $4n - 1$.

    Or: 1st term is 3, add 4 $(n - 1)$ times, therefore the nth term is $3 + 4(n - 1) = 3 + 4n - 4 = 4n - 1$.

  c If 137 is in the sequence, then $4n - 1 = 137$.

    Therefore $4n = 138$ and $n = 138 \div 4 = 34.5$

    This is not a whole number.

    Therefore 137 cannot be in the sequence.

Exam-style questions

1 The first four terms of a sequence are 3, 8, 13, 18. Find the 50th term and the nth term of this sequence.

2 The first five terms of a sequence are 1, 6, 11, 16, 21. Find the formula for the nth term.

3 The first four terms of a sequence are 2, 9, 16, 23.

  a Find the nth term of this sequence.

  b Show that 300 is not in this sequence.

4 a Write down the formula for the nth term of the sequence 1, 4, 9, 16, 25, … .

  b Hence, or otherwise, find the formula for the nth term of the sequence 4, 13, 28, 49, 76, … .

5 a Write down the formula for the nth term of the sequence 3, 5, 7, 9, … .

  b Hence, or otherwise, find the formula for the nth term of the sequence $1 \times 3 + 1$, $2 \times 5 + 1$, $3 \times 7 + 1$, … .

6 The three sequences below are linked. Write down the formula for the nth term of the sequence (a). Use the answer to write down the formula for sequence (b) and hence find the formula for sequence (c).

  a 1, 8, 27, 64, …

  b 2, 16, 54, 128, …

  c −1, 10, 45, …

16 Variation

● Direct proportion

- In direct proportion both variables change in the same way – either both getting larger or both getting smaller.
- Using symbols, direct proportion is written as $y \propto x$ or $y \propto x^2$.
- The formulae for these are $y = kx$ or $y = kx^2$, respectively, where k is a constant.

Examiner's tip

These are the most common direct proportions but you could also meet $y \propto x^3$ and $y \propto \sqrt{x}$.

Sample question

The value, $\$V$, of a diamond is proportional to the square of its mass, Wg. A diamond weighing 14 g is worth $\$490$.

1 a Find the value of a diamond weighing 40 g.

   b Find the mass of a diamond worth $\$6000$.

Answer

1 a $V \propto W^2$ or $V = kW^2$ b $6000 = 2.5 \times W^2$
  $490 = k \times 14^2$, $k = 2.5$ $W^2 = 2400$, so $W = 48.99$
  So $V = 2.5 \times 40^2$, i.e. $V = \$4000$

Now try this

1 If $y \propto x^3$ and $y = 24$ when $x = 2$, find y when $x = 4$.

2 If $y \propto x$ and $y = 8$ when $x = 20$, find x when $y = 36$.

3 A car uses 14 litres of fuel to travel 80 kilometres. How much fuel will it use to travel 250 kilometres?

4 When an object is dropped, the distance, d metres, which it falls in t seconds is proportional to t^2. If $d = 122.5$ metres when $t = 5$ seconds, calculate d when $t = 7$ seconds.

● Inverse proportion

- In inverse proportion when one variable increases the other variable decreases.
- Using symbols inverse proportion is written as $y \propto \dfrac{1}{x}$ or $y \propto \dfrac{1}{x^2}$.
- The formulae for these are $y = \dfrac{k}{x}$ or $y = \dfrac{k}{x^2}$ respectively, where k is a constant.

Examiner's tip

These are the most common inverse proportions, but you could also come across $y \propto \dfrac{1}{\sqrt{x}}$.

Sample question

The volume, V, of a given mass of gas varies inversely as the pressure, P. When $V = 2\,\text{m}^3$, $P = 500\,\text{N/m}^2$.

1 a Find the volume when the pressure is $400\,\text{N/m}^2$.

b Find the pressure when the volume is $5\,\text{m}^3$.

Answer

$1\ a\quad V \propto \dfrac{1}{P}$ or $V = \dfrac{k}{P}$

$2 = \dfrac{k}{500},\ k = 1000$

So $V = \dfrac{1{,}000}{400} = 2.5\,m^3$

$b\quad P = \dfrac{k}{V} = \dfrac{1{,}000}{5} = 200\,N/m^2$

Now try this

5 y is inversely proportional to x^2, and $y = 4$ when $x = 1$. Find y when $x = 2$.

6 y is inversely proportional to $\sqrt{x}$. $y = 1.2$ when $x = 100$.
 a Calculate y when $x = 4$. **b** Calculate x when $y = 3$.

7 The wavelength of a sound-wave, Wm, is inversely proportional to the frequency, FHz. A sound-wave with a frequency of 300 Hz has a wavelength of 1.1 m. Calculate the wavelength of a sound-wave with frequency of 660 Hz.

8 The brightness of a light varies inversely as the square of the distance from the source. A lamp has a brightness of 20 lumens at a distance of 1 metre. How bright is it at 10 metres?

● Sample exam question and answer

From a point h metres above sea level the distance, d kilometres, to the horizon is given by $d \propto \sqrt{h}$. When $h = 100$ metres, $d = 35$ kilometres. Find d when $h = 25$ metres.

$d \propto \sqrt{h}$ or $d = k\sqrt{h}$

$35 = k\sqrt{100},\ k = 3.5$

So $d = 3.5 \times \sqrt{25}$

$17.5\ kilometres$

Exam-style questions

1 The variable y is directly proportional to x^2. Given that $y = 75$ when $x = 5$, find the value of y when $x = 10$.

2 The number of coins, N, that can be made from a given volume of metal is given by $N \propto \dfrac{1}{d^2}$, where d cm is the diameter. Given that 8000 coins with a diameter of 2 cm can be made from the volume of metal, how many coins with a diameter of 4 cm can made from the same volume of metal?

3 The distance travelled by a car after the brakes are applied is proportional to the square of the initial speed. If it takes 12.5 metres to stop when travelling at $50\,\text{km}\,\text{h}^{-1}$ how far will a car travel if its initial speed is $120\,\text{km}\,\text{h}^{-1}$?

- To interpret and use graphs in practical situations, including travel graphs and conversion graphs.
- To draw graphs from given data.

- To apply the idea of rate of change to easy kinematics involving distance–time and speed–time graphs, acceleration and deceleration.
- To calculate distance travelled as area under a linear speed–time graph.

● Graphs in real situations

To interpret real-life graphs:

- look at the labels on the axes – they tell you what the graph is about
- notice whether the graph is a straight line or a curve
- the slope of the graph gives you the rate of change
- for distance–time graphs, the rate of change is the velocity
- for speed–time graphs, the rate of change is the acceleration
- the area under a speed–time graph = distance travelled.

Constant rate of change

Rate of change increasing

Rate of change decreasing

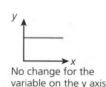

No change for the variable on the y axis

Examiner's tip

Remember to find the rate of change for each part of the graph. If it is zero, say how long this lasts.

Sample question

Asif walked to the bus stop, waited for the bus, and then travelled on the bus to school. Below is a distance–time graph for Asif's journey.

1 a How long did Asif wait at the bus stop?

b How far did Asif travel:
   (i) on foot
   (ii) by bus?

c How fast did Asif walk?

d What was the average speed of the bus?

e Why are the sections of the graph unlikely, in reality, to be totally straight?

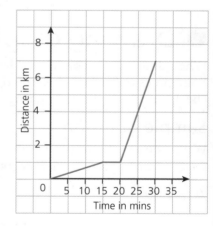

Answer

1 a 5 minutes
   b (i) 1 km
      (ii) 6 km

   c $\frac{1}{15}$ km/min or 4 km/h

d $\frac{6}{10}$ km/min or 36 km/h

e Stops and starts are likely to be gradual, not sudden.

Now try this

Describe what is really happening in these graphs.

1 a

(Graph showing Speed (m/s) on y-axis from 0 to 20, Time (secs) on x-axis from 0 to 50. Line rises from 0 to 20 at about 25 secs, stays constant to 40 secs, then drops to 0 at 50 secs.)

b

(Graph showing Volume of bath water (l) on y-axis from 0 to 40, Time (minutes) on x-axis from 0 to 25. Line rises from 0 to 20 at about 5 minutes, then to 40 at about 8 minutes, stays constant to 20 minutes, then drops to 0 at about 24 minutes.)

● Drawing graphs

When drawing a graph, always label your axes with the quantity and the unit. Sometimes only a sketch is asked for, not an accurate plot. You should always check whether the rate of change is constant, increasing or decreasing.

Sample question

Water is poured into this vessel at a constant rate. Sketch a graph of depth of water (d cm) against time (t seconds).

Answer

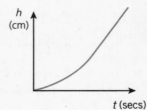

As the radius is decreasing at first, the rate of change will increase and then become constant.

Cambridge IGCSE Mathematics Study and Revision Guide Second Edition © Brian Seager *et al.*, 2016

Now try this

2 A car is travelling at 70 km h⁻¹, when it approaches some roadworks where the speed limit is 50 km h⁻¹. It slows down from 70 km h⁻¹ to 50 km h⁻¹ in 2 minutes. It goes through the roadworks at 50 km h⁻¹ in 5 minutes and then accelerates back to 70 km h⁻¹ in 30 seconds. Draw a velocity–time graph to illustrate this.

3 Find the distance travelled in the first 40 seconds in question 1(a).

4 Water is poured into this vessel at a constant rate. Sketch a graph of depth of water (*d* cm) against time (*t* seconds).

● Sample exam questions and answers

Tom leaves home at 8:20 a.m. and goes to school on a moped. The graph shows his distance from the school in kilometres.

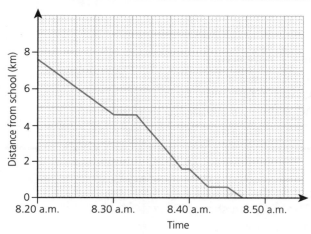

1 a How far does Tom live from school?

  b Write down the time that Tom arrives at the school.

  c Tom stopped three times on the journey. For how many minutes was he at the last stop?

  d Calculate his speed in kilometres per hour between 8:20 a.m. and 8:30 a.m.

1 a 7.6 km
  b 8:47 a.m.
  c 2.5 minutes
  d Distance = 7.6 − 4.6 = 3 km; time = 10 min; speed = $\frac{3}{10} \times 60 = 18$ km h⁻¹

Exam-style questions

1 Steve travelled from home to school by walking to a bus stop and then catching a school bus.

   a Use the information below to construct a distance–time graph for Steve's journey. Steve left home at 8:00 a.m. He walked at 6 km h^{-1} for 10 minutes. He then waited for 5 minutes before catching the bus. The bus took him a further 8 km to school at a steady speed of 32 km h^{-1}.

   b How far was Steve from home at 8:20 a.m?

2 The graph on the right describes a real-life situation. Describe what is possibly happening.

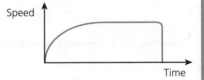

3 The diagrams show the cross-sections of three swimming pools. Water is pumped into all three at a constant rate. Sketch graphs of depth against time for each.

   a 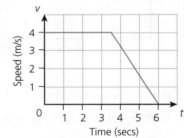   b   c

Area under a graph

To find the area under staight-line graphs, use the formulae for areas of a triangle, rectangle and trapezium. For the area under a curve, you need to estimate by counting squares or by dividing up the area into strips and approximating with trapezia. The area under a speed–time graph represents distance.

Exam-style questions

4 Ameni is cycling at 4 metres per second. After 3.5 seconds she starts to decelerate and after a further 2.5 seconds she stops. The diagram shows the speed–time graph of her journey. Calculate the distance travelled during the 6 seconds.

(18) Graphs of functions

Key objectives

- To construct tables of values for functions of the form $ax+b, +x^2\pm ax+b, \frac{a}{x}(x\neq 0)$, where a and b are integral constants.
- To draw and interpret such graphs.
- To solve linear and quadratic equations approximately by graphical methods.

- To construct tables of values and draw graphs for functions of the form ax^n, where a is a rational constant and $n = -2, -1, 0, 1, 2, 3$, and simple sums of not more than three of these, and for functions of the form a^x, where a is a positive integer.
- To solve associated equations approximately by graphical methods.
- To draw and interpret graphs representing exponential growth and decay problems.
- To estimate gradients of curves by drawing tangents.

● Drawing straight-line graphs

There are two main ways of writing equations of straight-line graphs: the type $y = 2x + 1$, and the type $3x + 2y = 6$.

To draw the first type:

Step 1: Draw up a table of venues using three x numbers.
Step 2: Substitute these into the formula to find the corresponding y values.
Step 3: Plot the coordinate pairs. They should lie in a straight line.

To draw the second type:

Step 1: Draw up a table of values. Use $x = 0$, $y = 0$ and one other x value.
Step 2: Substitute these into the formula to find the other values.
Step 3: Plot the coordinate pairs.

Sample question

Draw the line $y = 2x + 1$.

Answer

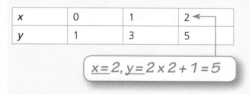

x	0	1	2
y	1	3	5

$\underline{x = 2, \ y = 2 \times 2 + 1 = 5}$

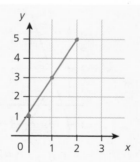

> **Now try this**
>
> Draw the following straight-line graphs
>
> 1 $y = 3x - 2$
> 2 $2x + 5y = 10$
> 3 $y = 6 - 2x$

Sample question

Draw the line $3x + 2y = 6$.

Answer

x	0	2	1
y	3	0	$1\frac{1}{2}$

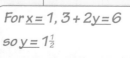

For $\underline{x = 0}$, $2y = 6$ so $\underline{y = 3}$

For $\underline{x = 1}$, $3 + 2y = 6$ so $\underline{y = 1\frac{1}{2}}$

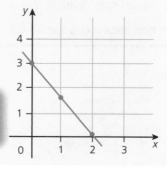

Cambridge IGCSE Mathematics Study and Revision Guide Second Edition © Brian Seager *et al.*, 2016

Graphical solution of simultaneous equations

Simultaneous equations can also be solved by graphical methods. The solution is found where the graphs intersect.

Step 1: Work out some values for each equation.

Step 2: Draw the two graphs on the same grid.

Step 3: Find where the two lines cross. Write down the x and y values of the point of intersection.

Sample question

Solve graphically $y = 4x - 1$ and $x + y = 4$.

Answer

The line $y = 4x - 1$ passes through $(0, -1)$ and $(2, 7)$, and $x + y = 4$ passes through $(0, 4)$ and $(4, 0)$.

Solution is $x = 1, y = 3$

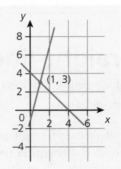

Now try this

4 Solve these simultaneous equations graphically
   a $y = 2x - 1, y = x + 2$
   b $y = 2x - 5, 2x + y = 3$

Sample exam question and answer

Solve these simultaneous equations graphically: $y = 3x - 1, y = 3 - 2x$

The line $y = 3x - 1$ passes through $(0, -1)$, $(1, 2)$ and $(2, 5)$.

The line $y = 3 - 2x$ passes through $(0, 3)$, $(1, 1)$ and $(3, -1)$.

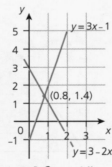

$x = 0.8, y = 1.4$

Exam-style questions

1 Solve these simultaneous equations graphically: $y = 3 - x, y = 3x - 2$.

2 Solve these simultaneous equations graphically: $y = 3x + 4, x + y = 2$.

● Quadratic graphs

Graphs of equations $y = ax^2 + bx + c$ are parabolas.
Their shape is

For $a > 0$ For $a < 0$

All parabolas are symmetrical. This symmetry can be seen on the graph and can often be seen in the table as well.

Sample question

Complete a table of values for $y = x^2 - 4x + 3$ and draw the graph for $-1 \leq x \leq 5$.

Answer

x	−1	0	1	2	3	4	5
y	8	3	0	−1	0	3	8

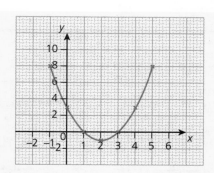

Now try this

5 a Complete this table for $y = x^2 - 3x + 4$.

x	−1	0	1	2	3	4
y						

   b Explain why it is a good idea to add an extra column to the table and work out the value of y when $x = 1.5$.
   c Draw the graph for $-1 < x < 4$.

6 Complete this table and draw the graph for $y = 2x^2 - 5x$, for $-2 \leq x \leq 5$.

x	−2	−1	0	1	2	3	4	5
y								

● Using graphs to solve quadratic equations

The solution of $ax^2 + bx + c = k$ is the values of x where the graph of $y = ax^2 + bx + c$ crosses the line $y = k$.
   The solution of the simultaneous equations $y = ax^2 + bx + c$ and $y = mx + k$ is the values of x where the two graphs cross.

Sample question

Use the graph of $y = x^2 - 4x + 3$ (drawn above) to solve the following equations.

1 a $x^2 - 4x + 3 = 0$ **b** $x^2 - 4x + 3 = 6$

  c $x^2 - 4x + 3 = x$ **d** $x^2 - 2x - 2 = 0$

Answer

1 a Look at where the graph crosses $y = 0$, the x-axis; the solution is $x = 1$ or 3.
  b Look at where the graph crosses the line $y = 6$; the solution is $x = -0.6$ or 4.6, to 1 d.p.
  c This is where the graph crosses the line $y = x$. When this line is added to the graph, it can be seen that the solution is $x = 0.7$ or 4.3, to 1 d.p.
  d Manipulating $x^2 - 2x - 2 = 0$ to give $x^2 - 4x + 3$ on the left-hand side gives $x^2 - 4x + 3 = 5 - 2x$.
   So the line to draw is $y = 5 - 2x$. When this line is added to the graph, it can be seen that the solution is $x = 2.7$ or -0.7, to 1 d.p.

Now try this

7 Use the graph you drew in 'Now try this' question 5 to solve the following equations
  a $x^2 - 3x + 4 = 4$ b $x^2 - 3x + 4 = 7$ c $x^2 - 3x + 4 = x + 2$

8 Use the graph you drew in 'Now try this' question 6 to solve the following equations
  a $2x^2 - 5x = 0$ b $2x^2 - 5x = -1$ c $2x^2 - 7x = 1$

9 Solve graphically, drawing the graphs for values of x from -1 to 3, these simultaneous equations: $y = x^2 - 2x + 1$, $y = 2 - x$.

● Cubic, reciprocal and exponential graphs

You also need to be able to recognise the shape of cubic, reciprocal and exponential graphs and to draw these graphs.

The cubic graph $y = ax^3$ has this shape when $a > 0$, and is reflected in the x-axis when $a < 0$.

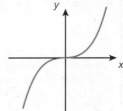

The reciprocal graph $y = \frac{a}{x}$ has this shape when $a > 0$, and is reflected in the x-axis when $a < 0$.

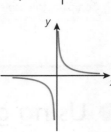

The exponential graph $y = k^x$, for $k > 1$, has this shape.

You will be asked to work out only the values for integer values of x.

All exponential graphs go through the point $(0, 1)$, as $k^0 = 1$ for all positive values of k. When $k > 1$, they increase steeply for $x > 0$, and are small when x is negative. This is reversed for $0 < k < 1$.

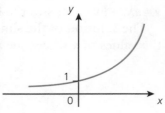

Sample question

Draw the graph of $y = x^3 - 3x^2$ for values of x from -1 to 4.

Use your graph to solve the equation $x^3 - 3x^2 = -1$.

Make a table for the values.

Answer

x	−2	−1	0	1	2	3	4
y	−20	−4	0	−2	−4	0	16

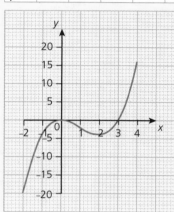

The solution of the equation is where the curve crosses the line $y = -1$.

The solution is $x = -0.5$, 0.7 or 2.9.

Examiner's tip

In this example the shape of the 'double bend' is more pronounced than in the graph of $y = ax^3$. This is because of the extra term $-3x^2$ in the equation.

Sample question

Draw $y = 3^x$ and $y = \left(\frac{1}{2}\right)^x$ on the same grid.

Answer

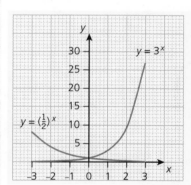

10 Draw the graph of $y = \frac{12}{x}$ for values of x from -12 to 12. Use the same scale on both axes.

11 Draw the graph of $y = x^3 - 2x$ for values of x from -2 to 2. Use your graph to find the roots of the equation $x^3 - 2x = 0$ to 1 d.p. Then use trial and improvement to find the positive root correct to 2 d.p.

12 Draw the graph of $y = 2^x$ for values of x from -1 to 4. Use your graph to find, to 1 d.p., the value of x for which $2^x = 10$.

Sample question

Plot the graph of $y = \dfrac{1}{x^2}$ for $-3 \leq x \leq 3$.

By drawing a suitable line, use your graph to solve the equation $2x^3 - 5x^2 - 1 = 0$.

Answer

To find the equation of the required line, take $2x^3 - 5x^2 - 1$ and manipulate it to get $\dfrac{1}{x^2}$.

$2x^3 - 5x^2 - 1 = 0$. Add 1 to both sides.

$2x^3 - 5x^2 = 1$. Factorise left-hand side.

$x^2(2x - 5) = 1$. Divide both sides by x^2.

$2x - 5 = \dfrac{1}{x^2}$

The equation of the required line is $y = 2x - 5$.

Plot the curve $y = \dfrac{1}{x^2}$ and the line $y = 2x - 5$.

Read off the x values where the curve and the line intersect.

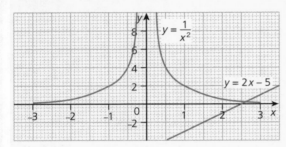

The solution of the equation $2x^3 - 5x^2 - 1 = 0$ from the graph is $x = 2.6$.

Now try this

13 Draw the graph of $y = x^2 - 3x + 5$ for $-4 \leq x \leq 5$.

Using the graph, and by drawing suitable lines, solve the following equations

   a $x^2 - 3x + 5 = 4$ **b** $x^2 - 4x + 2 = 0$ **c** $x^2 - 5x = 0$

14 Draw the graph of $y = \dfrac{1}{x}$ for $-5 \leq x \leq 5$. Use your graph to solve the following equations

   a $\dfrac{1}{x} = 3x + 1$ **b** $x^2 - 2x - 1 = 0$

15 Draw the graph of $y = \dfrac{1}{x^2}$ for $-3 \leq x \leq 3$. Use your graph to solve the equation $x^3 + 2x^2 - 1 = 0$.

● Sample exam question and answer

1 a Complete the table below for $y = x^3 - 2x^2 + 1$.

x	−1	−0.5	0
y		0.375	

   b Part of the graph is drawn on the grid. Add the three points from the table and complete the curve.

   c Use the graph to solve the equation $x^3 - 2x^2 + 1 = 0$.

   d By drawing a suitable straight line on the graph, solve the equation $x^3 - 2x^2 - x + 1 = 0$.

Examiner's tip

In this question, part of the graph was drawn.

This helped not only with drawing the graph but also with some of the answers for parts c and d.

Cambridge IGCSE Mathematics Study and Revision Guide Second Edition © Brian Seager *et al.*, 2016

1 a

x	−1	−0.5	0
y	−2	0.375	1

b

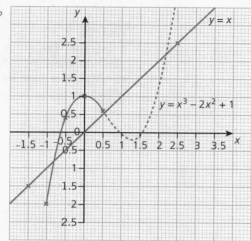

$y = x^3 - 2x^2 + 1$

$y = x$

c $x = -0.6, 1, 1.6$

d Manipulating the equation gives $x^3 - 2x^2 + 1 = x$. So the line to draw is $y = x$. The solution is $x = -0.8, 0.6$ or 2.2.

Exam-style questions

3 a Complete the table for $y = 4x - x^2$ and draw the graph.

x	−1	0	1	2	3	4	5
y			3			0	

b Use your graph to find

   (i) the value of x when $4x - x^2$ is as large as possible

   (ii) between which values of x the value of $4x - x^2 - 2$ is larger than 0.

4 The diagram shows the graphs P, Q, R, S, T and U. State which of these graphs could correspond to each of the following equations.

  a $y = x^3 - 1$ **b** $y = x^2 - 1$ **c** $y = x - 1$

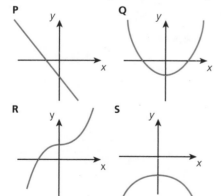

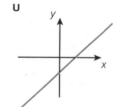

5 a Complete the table of values and draw the graph of $y = x^3 - 4x - 1$ for values of x from −3 to 3.

x	−3	−2	−1	0	1	2	3
y	−16		2	−1	−4	−1	14

b Use your graph to solve the equation $x^3 - 4x - 1 = 0$.

c By drawing a suitable straight line on your graph, solve the equation $x^3 - 6x - 3 = 0$.

6 Draw the graph of $y = x^3$ for $-2 \leq x \leq 2$. Use your graph to solve the equation $x^3 - 2x - 1 = 0$.

7 Draw the graph of $y = \dfrac{2}{x^2} - x$ for $-4 \leq x \leq 3$.

Use your graph to solve the equation $2x^3 + 7x^2 - 6 = 0$.

● Gradient

$$\text{Gradient} = \frac{\text{increase in } y}{\text{increase in } x}$$

The gradient of a curve varies at different points on the curve. To find the gradient of a curve at a given point, first draw a tangent to the curve at that point. Then find the gradient of the tangent. The gradient of the curve is the same as the gradient of the tangent, but because of inaccuracies in drawing can be regarded only as an estimate.

Examiner's tip

Remember a line that goes 'downhill' from left to right has a negative gradient.

Sample question

Find the gradient of the curve $y = x^2$ at the point $(2, 4)$.

Answer

Plot the graph and draw a tangent to the curve at (2, 4).

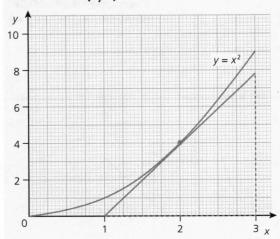

$$\text{Gradient} = \frac{7.8 - 0}{3 - 1} = \frac{7.8}{2} = 3.9.$$

Examiner's tip

When choosing two points on a tangent to find the gradient, make the x increase an easy number.

Now try this

15 Draw a graph of $y = x^3$ for $0 \le x \le 3$. Estimate the gradient at
  a (1, 1)   b (2, 8)

16 Draw a graph of $y = \dfrac{12}{x}$ for $-12 \le x \le 12$. Estimate the gradient at
  a $x = 2$ b $x = 4$ c $x = -4$

17 Draw a graph of $y = x^2 - 3x$ for $-2 \le x \le 5$. Estimate the gradient at
  a $x = -1$ b $x = 2$

18 Draw a graph of $y = x^3 - 2x$ for $-3 \le x \le 3$. Estimate the gradient at
  a $x = -2$ b $x = 0$ c $x = 2$

Cambridge IGCSE Mathematics Study and Revision Guide Second Edition © Brian Seager *et al.*, 2016

● Rates of change

- When the graph is a practical graph the gradient is called a rate of change and has units.
- The gradient of a distance–time graph represents speed (usual units $m\,s^{-1}$).
- The gradient of a speed–time graph represents acceleration (usual units $m\,s^{-2}$).

Sample question

The graph shows the area ($A\,cm^2$) of an ink blot against time (t seconds).
Find the rate of change when $t = 4$.

Answer

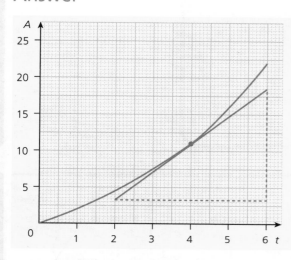

Draw a tangent to the curve at $t = 4$.

$$\text{Rate of change} = \frac{18.5 - 3.5}{6 - 2}$$

$$= \frac{15}{4}$$

$$= 3.75\,cm^2 s^{-1}.$$

Now try this

19 The graph shows distance (d m) against time (t secs). Estimate the speed at

  a $t = 2$ b $t = 4$

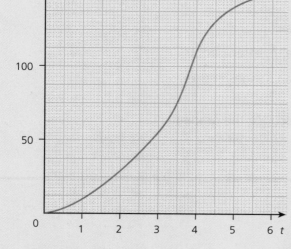

Now try this

20 The graph shows speed (v m s^{-1}) against time (t seconds). Estimate the acceleration at
   a $t = 2$ b $t = 8$

Exam-style questions

8 Ameni is cycling at 4 metres per second. After 3.5 seconds she starts to decelerate and after a further 2.5 seconds she stops. The diagram shows the speed–time graph of her journey. Calculate the acceleration between $t = 3.5$ and $t = 6$.

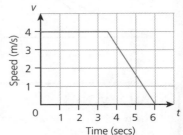

9 The number of bacteria in a colony multiplies by a factor of 3 every hour. Initially there are 20 bacteria.

   a Copy and complete the table

Time (t, hours)	0	1	2	3	4	5
Number of bacteria (n)	20	60				

   b Write down the formula for the number of bacteria, n, after t hours.

   c Draw a graph showing how the number of bacteria changes in the 5 hours.

   d (i) By drawing a tangent to the curve in part (c), calculate the gradient of the curve at the point where $t = 3.5$.

      (ii) What information does this gradient represent?

Functions

● Notation

$f: x \rightarrow 3x + 2$ means that f is the function that maps x on to $3x + 2$. It is more commonly written as: $f(x) = 3x + 2$. This is read as f of x equals $3x + 2$.

● Using functions

$f(7)$ means the value of the function when $x = 7$.

To find the value of a function, substitute the number or expression in the brackets for x in the formula.

When you are given the value of the function and want to find the value of x, put the formula for the function equal to the number or expression. Then solve the equation to find the value of x.

Sample questions

For the function $f(x) = 3x + 2$, find

1 a $f(7)$ **b** $f\left(\frac{1}{2}\right)$ **c** $f(2x)$

For the function $g(x) = 2x - 1$, find the value of x for which

2 a $g(x) = 11$ **b** $g(x) = x + 7$

Answers

1 a $f(7) = 3 \times 7 + 2 = 23$

  b $f\left(\frac{1}{2}\right) = 3 \times \frac{1}{2} + 2 = 3\frac{1}{2}$

  c $f(2x) = 3 \times (2x) + 2 = 6x + 2$

2 a $\quad g(x) = 11 \qquad$ b $\quad g(x) = x + 7$

      $2x - 1 = 11 \qquad\qquad 2x - 1 = x + 7$

         $2x = 12 \qquad\qquad 2x - x = 7 + 1$

          $x = 6 \qquad\qquad\qquad x = 8$

Now try this

1 For the function $f(x) = 5x + 1$, find
   a $f(7)$ b $f(-3)$ c $f(0)$

2 For the function $g(x) = 3 - 4x$, find
   a $g(6)$ b $g\left(1\frac{1}{2}\right)$ c $g(2x)$

3 For the function $h(x) = 3(2x - 7)$, find
   a $h(5)$ b $h(3x)$ c $h(x - 1)$

4 For the function $p(x) = \frac{2}{1 - x}$, find
   a $p(0)$ b $p(-4)$ c $p\left(\frac{1}{2}\right)$

5 For the function $q(x) = 3x^2 - 2$, find
   a $q(5)$ b $q(-1)$ c $q(2x)$ d $q(x + 3)$
   e $q(2x) + 5$

6 For the functions $f(x) = 5x - 4$ and $g(x) = 7x + 2$, find x for which
   a $f(x) = 6$ b $g(x) = -12$ c $f(x) = g(x)$ d $3f(x) = 2g(x)$

● Composite functions

When two or more functions are combined, the result is called a composite function.

The composite of the two functions f(x) and g(x) is written fg(x) or f[g(x)]. In general, the composite fg(x) is *not* the same as gf(x).

To find the value of a composite function fg(x), find the value of g(x) first, then put the answer in place of x in the formula for f(x).

Sample questions

For f(x)=$3x+2$ and g(x)=x^2-2x+1

1 find   **a** fg(3)   **b** gf(0)

2 find the formula for   **a** fg(x)   **b** gf(x)

Answers

1 a fg(3) = f[g(3)]
        = f[$3^2 - 2 \times 3 + 1$].
         Find g(3) first.
        = f[4] = $3 \times 4 + 2$. Then put
         the answer in place of x in
         the formula for f(x).
        = 14

 b gf(0) = g[f(0)]
        = g[$3 \times 0 + 2$]
        = g[2]
        = $2^2 - 2 \times 2 + 1$
        = 1

2 a fg(x) = f[g(x)]
        = f[$x^2 - 2x + 1$]. Replace x in
         the formula for f(x) with
         the formula for g(x).
        = $3(x^2 - 2x + 1) + 2$
        = $3x^2 - 6x + 5$

 b gf(x) = g[f(x)]
        = g[$3x + 2$]
        = $(3x + 2)^2 - 2(3x + 2) + 1$
        = $(9x^2 + 12x + 4) - 6x - 4 + 1$
        = $9x^2 + 6x + 1$

Replace every x in the formula for g(x) with the formula for f(x).

Now try this

7 For f(x)=$x-3$ and g(x)=$x+1$, find
  a fg(3)
  b gf(−3)
  c a formula for
    (i) fg(x)
    (ii) gf(x)

8 For h(x)=$5x+7$ and k(x)=$2x-4$, find
  a kh(4)
  b hk$\left(-\frac{1}{2}\right)$
  c a formula for
    (i) kh(x)
    (ii) hk(x)

9 For p(x)=$4x+7$ and q(x)=$\left(\frac{x+3}{2}\right)$, find
  a pq(5)
  b qp$\left(\frac{1}{2}\right)$
  c a formula for
    (i) pq(x)
    (ii) qp(x)

10 For f(x)=$1-3x$ and g(x)=$2(x+4)$ find
  a fg(0)
  b gf(−2)
  c a formula for
    (i) fg(x)
    (ii) gf(x)

11 For h(x)=$4x-3$ and j(x)=x^2 find a formula for
  a hj(x)
  b jh(x)

12 For f(x)=$1+2x$ and g(x)=x^2-2x+1 find a formula for
  a fg(x)
  b gf(x)

Cambridge IGCSE Mathematics Study and Revision Guide Second Edition © Brian Seager *et al.*, 2016

● Inverse functions

An inverse function has exactly the opposite effect to that of a given function.

For the function f(x), the inverse is written f^{-1}(x) so that when f(x)=x+1, f^{-1}(x)=x-1.

To find the formula of an inverse function, put y in place of f(x). Rearrange the formula to make x the subject. Replacing y with x gives the formula of the inverse function.

Sample question

For t(x)=5x-4, find
a a formula for t^{-1}(x)   **b** t^{-1}(1)   **c** t^{-1}(-3)

Answer

a $t(x) = 5x - 4$
$\quad y = 5x - 4$
$\quad 5x = y + 4$
$\quad x = \dfrac{y+4}{5}$, so $t^{-1}(x) = \dfrac{x+4}{5}$

b $t^{-1}(1) = \dfrac{1+4}{5} = 1$

c $t^{-1}(-3) = \dfrac{-3+4}{5} = \dfrac{1}{5}$

Alternative method for b and c
without finding $t^{-1}(x)$.

b $1 = 5x - 4$
$\quad x = 1$, so $t^{-1}(1) = 1$

c $-3 = 5x - 4$
$\quad x = \frac{1}{5}$, so $t^{-1}(-3) = \frac{1}{5}$

Now try this

13 For f(x)=7x+2, find
   a f^{-1}(x)   b f^{-1}(2)   c f^{-1}(0)

14 For g(x)=$\dfrac{x}{4}$-1, find
   a g^{-1}(x)   b g^{-1}$\left(1\frac{1}{4}\right)$ c g^{-1}(-5)

15 Find the inverse of each of the following functions
   a p(x)=3(x+2)   b h(x)=$\dfrac{2x-5}{4}$ c k(x)=4(3x-6)   d t(x)=5x^2+3, where $x \geqslant 0$

● Sample exam question and answer

For the function f(x)=px+q, f(5)=7 and f(1)=3. Find the values of p and q.

You are given the value of the function for two values of x so you can form a pair of simultaneous equations.

$f(5) = p \times 5 + q$ so $5p + q = 7$

$f(1) = p \times 1 + q$ so $p + q = 3$

Subtract the simultaneous equations: $4p = 4$, so $p = 1$.

Substitute the value of p: $1 + q = 3$, so $q = 2$.

Exam-style questions

1 f: $x \rightarrow 2x - 1$ and g: $x \rightarrow x^2 - 1$. Find, in their simplest form

   a $f^{-1}(x)$ **b** $gf(x)$

2 $f(x) = x^{\frac{1}{3}}$ and $g(x) = 2x^2 - 5$.

   a Find   **b** Find expressions for

     (i) $g(4)$ **(i)** $gf(x)$

     (ii) $fg(4)$ **(ii)** $f^{-1}(x)$

3 f: $x \rightarrow 3x - 5$ and g: $x \rightarrow x + 1$

   a Calculate $f(-1)$ **b** Find $f^{-1}(x)$

   c Find $fg(x)$. Give your answer in its simplest form

   d Solve the equation $3f(x) = 5g(x)$

4 $f(x) = \frac{1}{3}(2x + 5)$

   a Find $f(-4)$ **b** Find $f^{-1}(x)$

5 $f(x) = 3x + 1$ and $g(x) = 2x^2$. Find, in their simplest form

   a $f^{-1}(x)$ **b** $fg(2)$ **c** $gf(x)$ **d** $ff(x)$

6 $f(x) = 2x + 1$ and $g(x) = \frac{x}{3} - 5$

   a Find $f(-1)$ **b** Find $f^{-1}(x)$ **c** Find $g^{-1}(x)$ **d** Find $fg(0)$

   e Find $fg(x)$ in its simplest form

Cambridge IGCSE Mathematics Study and Revision Guide Second Edition © Brian Seager *et al.*, 2016

20 Geometrical vocabulary

● Angles and lines

- Acute angles are angles between 0° and 90°.
- Right angles are 90°.
- Obtuse angles are between 90° and 180°.
- Reflex angles are between 180° and 360°.
- Perpendicular lines meet at a right angle.
- Parallel lines never meet and are marked with arrows.

● Triangles

- Isosceles triangles have two equal sides and two equal angles.
- Equilateral triangles have three equal sides and three equal angles.
- Scalene triangles have all three sides with different lengths.
- In a right-angled triangle the longest side is called the hypotenuse.

● Congruent shapes

Congruent shapes are identical. All the corresponding angles are equal and all the corresponding sides are equal.

For two triangles to be congruent there are four possible conditions:

- All the corresponding sides are equal (SSS).
- Two corresponding sides are equal and the angle between them is the same (SAS).
- Two angles and a corresponding side are equal (AAS).
- The triangles are right angled and have their hypotenuses and a corresponding side equal (RHS).

● Similar shapes

Two shapes are similar if they have all their corresponding angles equal and all their corresponding sides proportional. In the case of a triangle, if one of these facts is true, then so is the other.

Examiner's tip

So it is enough to know that either all the angles are equal or all the sides are proportional.

● Circles

The names connected with circles are shown in these two diagrams.

The segment in the right-hand diagram is called a minor segment. A segment that is more than half the circle is called a major segment.

The sector in the right-hand diagram is called a minor sector. A sector that is more than half the circle is called a major sector.

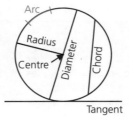

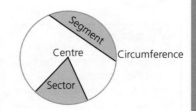

● Quadrilaterals

The table below summarises the special quadrilaterals.

Shape	Name	Sides	Angles	Diagonals
■	Square	All equal	All equal	Equal and bisect each other at right angles
▬	Rectangle	Opposite sides equal	All equal	Equal and bisect each other
◆	Rhombus	All equal	Opposite angles equal	Bisect each other at right angles
▱	Parallelogram	Opposite sides equal	Opposite angles equal	Bisect each other
▱	Trapezium	One pair of opposite sides parallel		
◆	Kite	Two pairs of adjacent sides equal	One pair of opposite angles equal	Intersect at right angles

● Polygons

Polygons are shapes with straight sides. The table below gives the numbers of sides and names.

3	4	5	6	7	8	10	12
Triangle	Quadrilateral	Pentagon	Hexagon	Heptagon	Octagon	Decagon	Dodecagon

● Nets of solids

The net of a solid is the shape you would cut out and fold to make the solid.

This is a possible net for a cube.

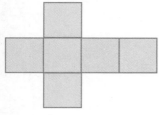

Cambridge IGCSE Mathematics Study and Revision Guide Second Edition © Brian Seager *et al.*, 2016

Sample questions

For this accurately drawn diagram write down

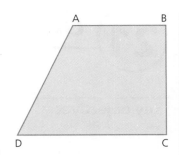

1 an acute angle   3 two lines perpendicular to each other

2 an obtuse angle   4 two parallel lines

Answers

1 angle D   3 AB and BC or DC and BC
2 angle A   4 AB and DC

● Sample exam question and answer

From this list of words pick the right one for each label on the diagram

Circumference
Tangent
Chord
Sector
Radius
Segment
Arc
Diameter

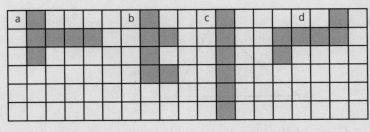

a Diameter   c chord
b arc   d radius

Now try this

1 State whether these angles are acute, obtuse or reflex:
  a   63°   c   94°   e 339°
  b 248°   d 168°
2 Which of these are possible nets for a cube?

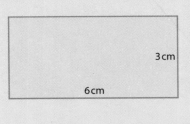

3 Explain why these two rectangles are NOT similar. (Rectangles not to scale.)

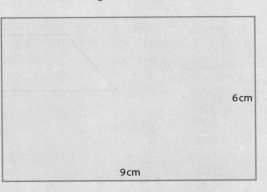

3 cm

6 cm

6 cm

9 cm

Cambridge IGCSE Mathematics Study and Revision Guide Second Edition © Brian Seager *et al.*, 2016

Geometrical constructions and scale drawings

Key objectives

- To measure lines and angles.
- To construct a triangle given the three sides using only a ruler and a pair of compasses.
- To construct other simple geometrical figures from given data using a ruler and protractor as necessary.
- To construct angle bisectors and perpendicular bisectors using only a straight edge and a pair of compasses.
- To read and make scale drawings.

● Drawing shapes

You should be able to construct a triangle given three sides using only a ruler and compasses. You should be able to construct other shapes using a ruler, compasses and protractor.

Triangle with three sides given

Step 1: Draw line AB of given length.
Step 2: Use compasses to construct arcs AC and BC with the compasses set to the given lengths.
Step 3: Draw AC and BC.

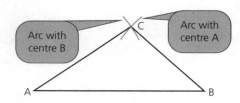

Other shapes

Triangle knowing two sides and the included angle

Step 1: Draw line AB of given length.
Step 2: Measure and draw angle at A.
Step 3: Draw line AC of given length.
Step 4: Join C to B.

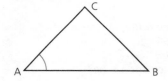

Triangle knowing angles and one side

Step 1: Draw line AB of given length.
Step 2: Measure and draw angles at A and B.
Step 3: Draw lines AC and BC.

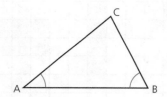

Parallelogram knowing two adjacent sides and the angle between them

Step 1: Draw line AB of given length.
Step 2: Measure and draw angle at A.
Step 3: Measure and draw the same angle at B.
Step 4: Measure sides AD and BC.
Step 5: Draw the line DC.

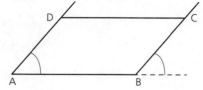

Cambridge IGCSE Mathematics Study and Revision Guide Second Edition © Brian Seager *et al.*, 2016

Now try this

1 Draw these triangles accurately.
  For each triangle measure the length of any unmarked sides and the size of any unknown angles.

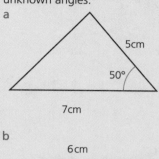

a

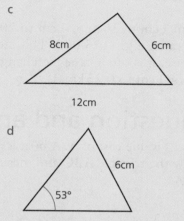

c

b

d

Angle constructions

You also need to construct, using a ruler and compasses, (1) the perpendicular bisector of a line and (2) the bisector of an angle.

Example 1

AB is the given line. With centres A and B draw arcs of the same radius, as shown. Join the intersections of the arcs. This line is the perpendicular bisector of line AB.

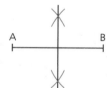

Example 2

AB and AC are the two lines making the given angle. With centre A draw an arc on AC and AB as shown. At these points draw two more arcs. Join the point where these arcs intersect to A. This line is the bisector of angle A.

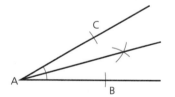

Now try this

2 Draw this triangle accurately. PX is the perpendicular bisector of line AB. AZ is the bisector of the angle at A and crosses PX at Z. Draw these lines accurately and measure the distance AZ.

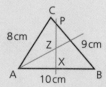

3 Draw, using ruler and compasses only, an equilateral triangle, ABC, of side 10 cm. Bisect two of the angles as shown, making an isosceles triangle, BXC. Measure the two equal sides, BX and CX, of this isosceles triangle.

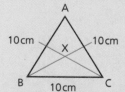

● Scale drawings

You may be given a scale in the form 1 cm to 10 m or 1 cm to 20 km.

Sometimes the scale is given as a ratio, e.g. 1 : 25 000. This means that the scale is 1 cm to 25 000 cm. In this case, it is often easier to change the units: 1 cm to 25 000 cm = 1 cm to 250 m or 1 cm to 0.25 km.

● Sample exam question and answer

Two buoys are anchored at A and B. B is due east of A. A boat is anchored at C.

Using a scale of 1 cm to 2 m, draw the triangle ABC and measure the bearing of the boat, C, from buoy A.

This is a standard construction as shown.

Draw line AB 7.5 cm long. Using compasses, measure an arc 10 cm from A and 4 cm from B. Mark the point C where they cross.

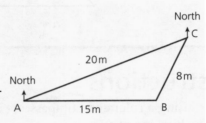

Using a protractor measure CAB and subtract the measurement from 90°. The bearing should be between 20° and 21°.

Now try this

4   A triangular field has sides 380 m, 250 m and 190 m. Using a scale of 1 : 5000, make an accurate scale drawing of the field.

Cambridge IGCSE Mathematics Study and Revision Guide Second Edition © Brian Seager *et al.*, 2016

(22) Similarity

Similar shapes

For shapes to be similar:

- The corresponding angles are equal.
- The corresponding sides are proportional.

In the special case of triangles, one of these conditions is sufficient. If one is true the other one must be.

If angle A = angle R, angle B = angle S and angle C = angle T, then the triangles are similar and it follows that:

$$\frac{AB}{RS} = \frac{BC}{ST} = \frac{AC}{RT} = k \text{ (the scale factor)}.$$

Sample question

1 a Prove that triangle ABC is similar to triangle PQR.
  b Find the length of PQ.

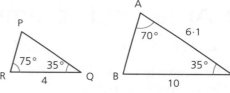

Answer

1 a Angle ABC = 180 − 70 − 35 = 75
    Angle RPQ = 180 − 75 − 35 = 70
    Triangles are similar as
    corresponding angles are equal.

  b $\frac{PQ}{4} = \frac{6.1}{10}$

    $PQ = \frac{6.1 \times 4}{10} = 2.4$ (1 d.p.)

Now try this

1 Triangles LMN and PQR are similar. Find the lengths of the missing sides.

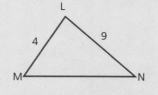 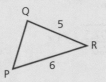

Now try this

2 The quadrilateral PQRS is similar to the quadrilateral ABCD. Find
   a   angle S   b   SR   c   BC

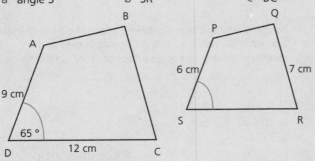

3 Triangle ABC is similar to triangle ADE. AB = 5 cm, BD = 3 cm and AC = 7 cm. Calculate the length of CE.

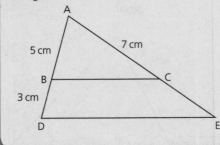

● Areas and volumes of similar shapes

If two shapes A and B are similar and the scale factor for lengths is k:

- The area scale factor = k².
- The volume scale factor = k³.

Sample question

Two similar cylinders have heights of 8 cm and 16 cm.

1 a The smaller has a volume of 60 cm³. Find the volume of the larger.

  b Another similar cylinder has a volume of 202.5 cm³. Find its height.

Answer

1 a Length scale factor $= \dfrac{16}{8} = 2$. Volume $= 60 \times 2^3 = 480\ cm^3$.

  b Volume scale factor $= \dfrac{202.5}{60}$. Length scale factor $= \sqrt[3]{\dfrac{202.5}{60}} = 1.5$.
     Height $= 8 \times 1.5 = 12\ cm$.

Cambridge IGCSE Mathematics Study and Revision Guide Second Edition © Brian Seager *et al.*, 2016

Now try this

4 In question 2 the area of quadrilateral ABCD is 76.5 cm³. Find the area of quadrilateral PQRS.

5 A solid metal sphere with a volume of 12 800 cm³ is melted down and made into smaller spheres each with $\frac{1}{8}$ the radius of the original sphere. Calculate the volume of each of the smaller spheres.

6 A toy van is made to a scale 1 : 50.
   a The actual van is 2.4 m wide. How wide is the model?
   b The front window of the model has an area of 4 cm². What is the window area of the actual van?

7 Two similar objects have volumes of 24 cm³ and 81 cm³. The surface area of the larger object is 540 cm². Find the surface area of the smaller object.

● Sample exam questions and answers

Two similar solids have heights of 9 cm and 12 cm.

1 The smaller solid has a surface area of 486 cm². Find the surface area of the larger solid.

2 The larger solid has a volume of 3200 cm³. Find the volume of the smaller solid.

1 Length scale factor $= \frac{12}{9} = \frac{4}{3}$

   Area scale factor $= \left(\frac{4}{3}\right)^2 = \frac{16}{9}$

   Area of larger solid $= 486 \times \frac{16}{9} = 864$ cm²

2 Length scale factor $= \frac{3}{4}$

   Volume scale factor $= \left(\frac{3}{4}\right)^3 = \frac{27}{64}$

   Volume of smaller solid $= 3200 \times \frac{27}{64} = 1350$ cm³

Key objectives

- To recognise rotational and line symmetry (including order of rotational symmetry) in two dimensions.
- To recognise the symmetry properties of the prism (including the cylinder) and the pyramid (including the cone).

- To use the following symmetry properties of circles:
  - equal chords are equidistant from the centre
  - the perpendicular bisector of a chord passes through the centre
  - tangents from an external point are equal in length.

● Two dimensions

A shape has line symmetry if it fits onto itself when it is reflected in a mirror line or folded along the mirror line.

A shape has rotational symmetry if it fits onto itself when it is turned round a central point.

The number of times it fits onto itself during a complete revolution is called the **order** of rotational symmetry.

Sample questions

For each of these shapes

1 find the number of lines of symmetry,

2 find the order of rotational symmetry.

a   b   c

Answers

a   1   One line of symmetry.
     2   Order of rotational symmetry = 1.
(This is the same as saying that there is no rotational symmetry.)

b   1   No lines of symmetry.
     2   Order of rotational symmetry = 2.

c   1   Four lines of symmetry.
     2   Order of rotational symmetry = 4.

Now try this

Copy each of these shapes.

1 Draw in all the lines of symmetry.
2 State the order of rotational symmetry.

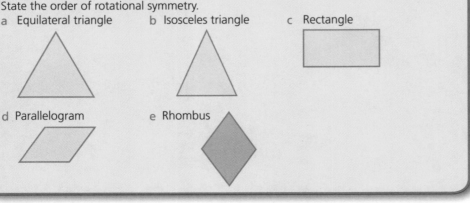

a Equilateral triangle   b Isosceles triangle   c Rectangle

d Parallelogram   e Rhombus

Cambridge IGCSE Mathematics Study and Revision Guide Second Edition © Brian Seager *et al.*, 2016

Three dimensions

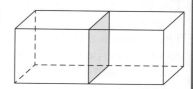

A solid has a plane of symmetry if it fits onto itself when it is reflected in the plane.

A solid has an axis of rotational symmetry if it fits onto itself when turned about a central axis.

The number of times it fits onto itself during a complete revolution is called the **order** of rotational symmetry about that axis.

Sample questions

For this cuboid with a square top

1 State the number of planes of symmetry.

2 State the number of axes of rotational symmetry and the order of symmetry for each one.

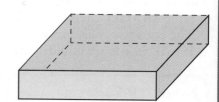

Answers

1 Five (three through the middle of opposite faces and two diagonal).
2 Three (one of order 4 and two of order 2).

Now try this

3 For these shapes
  a State the number of planes of symmetry.
  b State the number of axes of rotational symmetry and the order of symmetry for each one.

Equilateral triangular prism.

Square-based pyramid.

Circles, tangents and chords

Tangent properties of circles

The tangent at any point on a circle is at right angles to the radius at that point. The two tangents from a point outside a circle to the circle are equal in length.

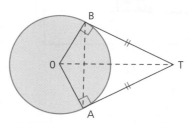

Figure A So angle TAO = angle TBO = 90°, and TA = TB.

Sample questions

In Figure A, angle ATB = 50°.
Calculate:
1 angle TAB;
2 angle AOB.

Answers

1 As TA = TB, triangle TAB is isosceles.
Angle TAB = $\frac{1}{2}(180° - 50°) = 65°$.

2 OT is a line of symmetry, so angle BTO = $\frac{1}{2}(50) = 25°$.
So angle BOT = 90 − 25 = 65°.
Angle AOB - 2 × 65 = 130°.

Chord properties of circles

The perpendicular bisector of a chord passes through the centre of the circle (Figure B).
The reason is that OA and OB are both radii, so triangle OAB is isosceles.
As ON is perpendicular to AB, it is the line of symmetry for the triangle.
Therefore AO = BO and O is the centre.

Similarly:

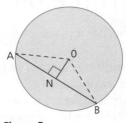

Figure B

- a line through the centre perpendicular to the chord bisects the chord and
- a line joining the centre of the circle to the midpoint of the chord is perpendicular to the chord.

Two equal chords are equidistant from the centre of the circle (Figure C).
So if AB = CD, then OP = OQ.
The converse is also true.
If two chords are equidistant from the centre, they are equal in length.

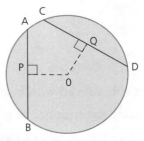

Figure C

Now try this

4 In Figure A, if angle ABO = 25°, find angle ATB.

● Sample exam question and answer

Shade four more squares so that the dashed lines are lines of symmetry of the pattern.

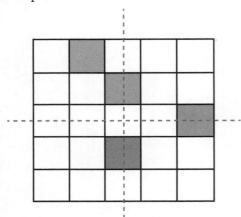

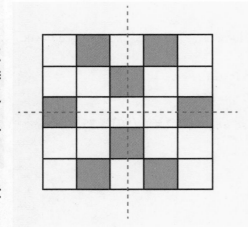

Key objectives

To calculate unknown angles using the following geometrical properties:

- angles at a point
- angles at a point on a straight line and intersecting straight lines
- angles formed within parallel lines
- angle properties of triangles and quadrilaterals
- angle properties of regular polygons

- angles in a semi-circle
- the angle between tangent and radius of a circle
- angle properties of irregular polygons
- the fact that the angle at the centre of a circle is twice the angle at the circumference
- the fact that angles in the same segment are equal
- the fact that angles in opposite segments are supplementary, cyclic quadrilaterals.

● Basic angle facts

The sum of the angles on a straight line is $180°$: $a + b + c = 180°$.

The sum of the angles round a point is $360°$: $a + b + c + d + e = 360°$.

When two lines cross, the opposite angles are equal. This is written as vertically opposite angles: $a = c$ and $b = d$.

The angles of a triangle add up to $180°$: $a + b + c = 180°$.

The exterior angle of any triangle is equal to the sum of the opposite interior angles. In the diagram d is the exterior angle: $d = b + c$.

● Angles with parallel lines

A line crossing a pair of parallel lines is called a transversal.

a and b are called alternate angles. They are on opposite sides of the transversal.

c and d are called corresponding angles. They are in the same position between the transversal and the parallel lines.

e and f are called co-interior angles (or sometimes interior or allied angles). They are the same side of the transversal between the parallel lines.

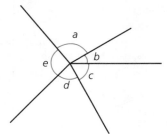

$a = b$

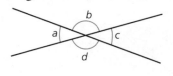

$c = d$

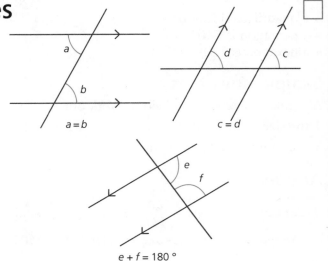

$e + f = 180°$

Sample question

Calculate the size of the angles marked with letters. Give reasons for your answers.

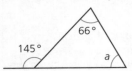

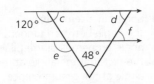

Answer

$a = 79°$ (exterior angle = sum of opposite interior angles)

$c = 60°$ (angles on a straight line = 180°)
$d = 72°$ (angles in a triangle = 180°)
$e = 120°$ (corresponding angles are equal)
$f = 72°$ (alternate angles are equal)

Now try this

Calculate the size of the angles marked with letters in questions 1 to 3. Give reasons for your answers.

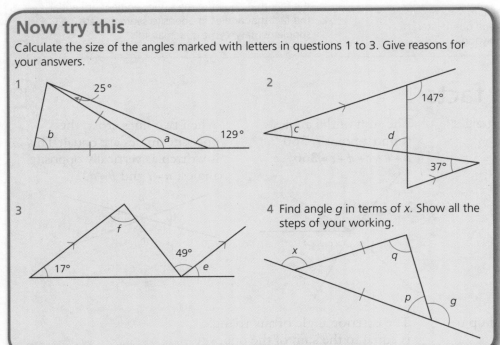

4 Find angle g in terms of x. Show all the steps of your working.

● Polygons

The sum of the exterior angles of a polygon is $360°$: $a + b + c + d + e + f = 360$.

The sum of interior angles of a polygon is equal to $180(n-2)°$.

For this hexagon $p + q + r + s + t + u = 180(6-2) = 720°$.

A regular polygon has all its sides and all its angles equal.

Often it is easier to use the exterior angles of a polygon rather than the interior angles.

It is useful to know that the sum of the interior angles of:

- a quadrilateral is $360°$
- a pentagon is $540°$
- a hexagon is $720°$.

Sample questions

A regular polygon has 12 sides. Work out

1 the size of each exterior angle   2 the size of each interior angle

3 the sum of the interior angles.

Answers

1 Each exterior angle $= \dfrac{360}{12} = 30°$.   2 Each interior angle $= 180 - 30 = 150°$.

3 The sum of the interior angles $= 150 \times 12 = 1800°$. Or $180(12-2) = 180 \times 10 = 1800°$.

Cambridge IGCSE Mathematics Study and Revision Guide Second Edition © Brian Seager *et al.*, 2016

Now try this

5 Find the interior angle of a regular octagon.
6 Three interior angles of a quadrilateral are 112°, 65° and 79°. Find the size of the fourth angle.
7 A regular polygon has an interior angle of 156°. How many sides has the polygon?
8 A pentagon has four interior angles of 75°, 96°, 125° and 142°. Find the size of the fifth interior angle.

● Angle properties of circles

The angle subtended by an arc at the centre is twice the angle subtended at the circumference.

Angle AOB = 2 × angle APB.
Reflex angle AOB = 2 × angle AQB.

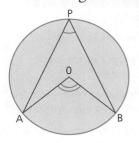

1

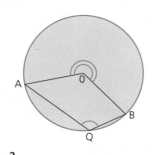

2

Sample questions

1 Angle AOB = 68°. Find angle APB.

2 Angle AQB = 115°. Find
a reflex angle AOB
b angle AOB.

The angle in a semi-circle is a right angle.

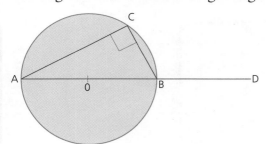

3

3 Angle CAO = 38° and O is the centre of the circle. AOBD is a straight line. Find angle CBD.

Angle ACB = 90°.

Angles in the same segment are equal.

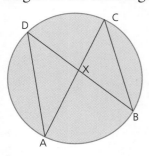

4

Angle ADB = angle ACB and angle DAC = angle DBC.

4 Angle ACB = 35°. Find angle ADB. Give your reasons.

Angles in the opposite segment add up to 180°.
   This property is also known as 'opposite angles of a cyclic quadrilateral'.

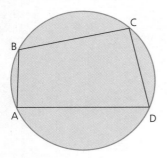

Angle ABC + angle CDA = 180°
and angle BAD + angle BCD = 180°.

5 Angle BCD = 102°. Find angle BAD. Give your reasons.

Answers

1 Angle APB = ½ angle AOB = ½ × 68° = 34°.

2 a Reflex angle AOB = 2 × angle AQB = 2 × 115° = 230°.
  b Angle AOB = 360° − 230° = 130°.

3 Angle ACB = 90°.
  So angle CBA = 180° − 38° − 90° = 52°.
  So angle CBD = 180° − 52° = 128°.

4 Angle ADB = 35°, because angles ACB and ADB are angles in the same segment.

5 Angle BAD = 180° − 102° = 78° because angles BCD and BAD are opposite angles of a cyclic quadrilateral.

Now try this

9 Find
  a angle CAD   b angle CBD

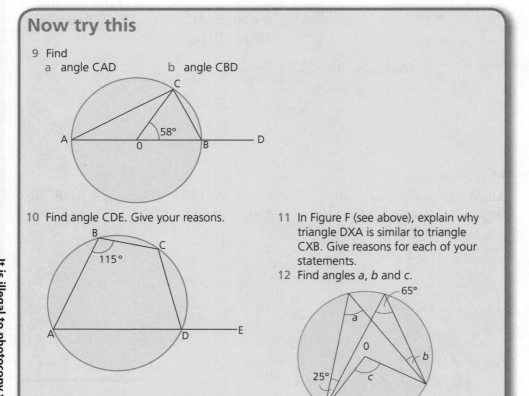

10 Find angle CDE. Give your reasons.

11 In Figure F (see above), explain why triangle DXA is similar to triangle CXB. Give reasons for each of your statements.

12 Find angles a, b and c.

Cambridge IGCSE Mathematics Study and Revision Guide Second Edition © Brian Seager *et al.*, 2016

Sample exam questions and answers

The sketch shows a regular pentagon and a regular hexagon with equal length sides.

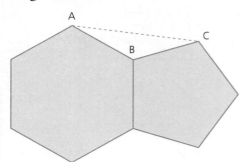

1 a Work out the size of one interior angle of
   (i) the pentagon
   (ii) the hexagon.

   b Work out the sizes of the angles in triangle ABC.

1 a (i) Exterior angle = $\frac{360}{5}$ = 72°. Interior angle = 180 − 72 = 108°.

   (ii) Exterior angle = $\frac{360}{6}$ = 60°. Interior angle = 180 − 60 = 120°.

   b In triangle ABC, angle ABC = 360 − 120 − 108 = 132°.
   Triangle ABC is isosceles, as AB = BC.

   So angles BAC and BCA = $\frac{180 - 132}{2} = \frac{48}{2}$ = 24°.

Sample exam questions and answers

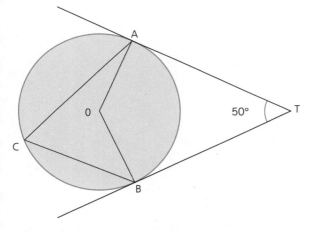

O is the centre of the circle and angle ATB is 50°. TA and TB are tangents.

1 Find angle AOB.

2 Find angle ACB, giving a reason for your answer.

1 Angle OAT = angle OBT = 90°. Angle AOB = 360 − 90 − 90 − 50 = 130°.

2 Angle ACB = $\frac{1}{2}$ × 130 = 65°. Because angle at centre = twice angle at circumference.

Key objectives

To use the following loci and the method of intersecting loci for sets of points in two dimensions that are:

• at a given distance from a given point

• at a given distance from a given straight line
• equidistant from two given points
• equidistant from two given intersecting straight lines.

● Basic loci

A locus is a path or region in which a point can move according to a rule.

The locus of a point that moves so it is 3 cm from a fixed point A is a circle, centre A, radius 3 cm.

The locus of a point that moves so it is less than 3 cm from a fixed point A is the region inside a circle, centre A, radius 3 cm.

The locus of a point that moves so it is more than 3 cm from a fixed point A is the region outside a circle, centre A, radius 3 cm.

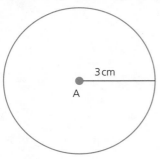

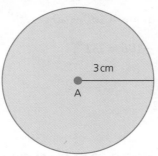

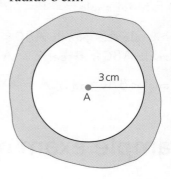

The locus of a point that stays 2 cm from a fixed straight line is one of a pair of lines parallel to that line.

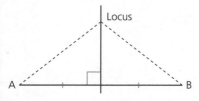

The locus of a point that stays an equal distance from two fixed points is the perpendicular bisector of the line joining the two points.

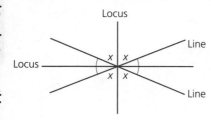

The locus of a point that stays an equal distance from two fixed lines is one or both of the angle bisectors of the lines.

Cambridge IGCSE Mathematics Study and Revision Guide Second Edition © Brian Seager *et al.*, 2016

Sample question

A goat is tethered by a 2-metre rope to a rail 3 metres long fixed in a field of grass. Draw a scale diagram to show the region of grass that the goat can eat.

Answer

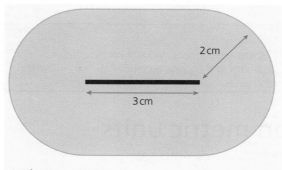

Scale: 1 cm to 1 metre.

Now try this

1 Draw the locus of a point that moves so it is 4 cm from a fixed point P.
2 A and B are fixed points 6.5 cm apart.
  a Draw the locus of points that are equidistant from A and B.
  b Shade the locus of points that are nearer A than B.
3 Two sides of a field meet at an angle of 70°. A footpath starts from this corner of the field and goes across the field keeping the same distance from each side of the field. Draw the footpath.
4 A stretch of coast has the shoreline running south-east to north-west. A boat moves keeping 500 metres from this shore. Draw a scale diagram to show its path.

● Problems involving intersecting loci

Often, exam questions involve more than one locus. Solve these questions in stages, drawing one locus at a time.

If the question does not tell you what to shade, you may either shade the required region or shade those not required. Shading those not required is often easier if there are several loci.

Examiner's tip

Remember to give a key or label the diagram to make it clear whether you have shaded the region required or not required.

● Sample exam question and answer

This is the plan of a garden drawn on a scale of 1 cm to 2 metres. A pond is to be dug in the garden. It must be at least 4 metres from the tree. It must be at least 3 metres from the house. Shade the region where the pond could be dug. Show all your construction lines.

Scale 1 cm to 2 metres. At least 4 metres from the tree — so outside a circle of radius 2 cm centred on the tree. At least 3 metres from the house, so to the left of a line parallel to the house and 1.5 cm from it. The possible position for the pond is shaded in red.

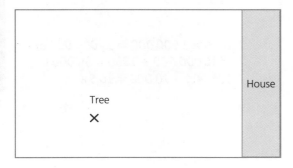

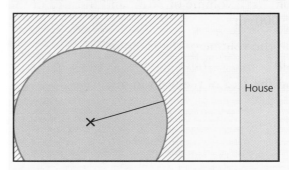

(26) Measures

Key objectives
- To use current units of mass, length, area, volume and capacity in practical situations and express quantities in terms of larger or smaller units.

● Common metric units

The most commonly used metric units are:

- Length: kilometre (km), metre (m), centimetre (cm), millimetre (mm).
- Mass: tonne (t), kilogram (kg), gram (g), milligram (mg).
- Capacity: litre (l), millilitre (ml).

● Changing units

Length	Mass	Capacity
1 cm = 10 mm	1 g = 1000 mg	1 ℓ = 1000 mℓ
1 m = 1000 mm	1 kg = 1000 g	1 ℓ = 1000 cm³
1 m = 100 cm	1 t = 1000 kg	1 m³ = 1000 ℓ (litres)
1 km = 1000 m		

Now try this
Change these units

1 40 mm to cm
2 150 cm to m
3 1.5 kg to g
4 52 ℓ to mℓ
5 15.3 cm to mm
6 1465 g to kg
7 1340 mℓ to ℓ
8 75 mm to m

● Area and volume

Area units are squares with sides equal to the length units, for example square metres, which is written as m^2.

Volume units are cubes with sides equal to the length units, for example cubic centimetres, which is written as cm^3.

● Changing units

Area	Volume
1 cm² = 10² mm² = 100 mm²	1 cm³ = 10³ mm³ = 1000 mm³
1 m² = 100² cm² = 10 000 cm²	1 m³ = 100³ cm³ = 1 000 000 cm³
1 km² = 1000² m² = 1 000 000 m²	

Now try this
Change these units

9   40 cm² to mm²
10 1.50 m² to cm²
11 1.5 m³ to ℓ
12 5 254 000 cm³ to m³
13 500 mm² to cm²
14 15.3 cm³ to mm³
15 6.3 m³ to cm³
16 24 000 ℓ to m³

● Sample exam questions and answers

A room has a volume of 36 m³ and the area of the walls is 105 000 cm².

1 Find the volume of the room in
  a cm³   b litres

2 Find the area of the walls in m².

1 a $36 \times 1\,000\,000 = 36\,000\,000\ cm^3$
  b $36\,000\,000 \div 1000 = 36\,000\ l$
2 $105\,000 \div 10\,000 = 10.5\ m^2$

Cambridge IGCSE Mathematics Study and Revision Guide Second Edition © Brian Seager *et al.*, 2016

(27) Perimeter, area and volume

Key objectives

- To carry out calculations involving the perimeter and area of a rectangle, triangle, parallelogram and trapezium and compound shapes derived from these.
- To carry out calculations involving the circumference and area of a circle.
- To be able to carry out calculations involving the volume of a cuboid, prism and cylinder and the surface area of a cuboid and a cylinder.
- To solve problems involving the arc length and sector area as fractions of the circumference and area of a circle.
- To be able to carry out calculations involving the surface area and volume of a sphere, pyramid and cone.
- To be able to carry out calculations involving the areas and volumes of compound shapes.

● Perimeter and area of two-dimensional shapes

Perimeter is the distance round the outside of a shape. Add up all the lengths of the sides.

You need to know the following formulae.

Rectangle: $A = h \times b$; $P = 2b + 2h$

Triangle: $A = \dfrac{b \times h}{2}$

Parallelogram: $A = b \times h$

Trapezium: $A = \dfrac{h}{2}(a + b)$

Circle: $A = \pi \times r^2$

The circumference is another name for the perimeter of a circle: $C = \pi \times d = 2\pi r$.

Sample questions

1 The vertices of a triangle are (2, 1), (2, −3) and (14, −3). Find its area.

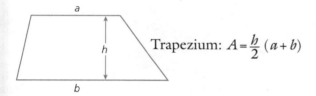

2 Find the area and perimeter of this shape made from a semicircle and a parallelogram.

3 A circular pond has an area of 40 m². Find the radius of the pond.

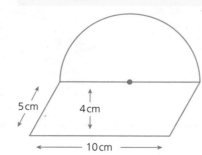

Answers

1 $A = \dfrac{12 \times 4}{2} = 24 \text{ units}^2$

2 $A = \frac{1}{2} \times \pi \times 5^2 + 10 \times 4$
$\quad = 39.3 + 40$
$\quad = 79.3 \text{ cm}^2$
$P = 5 + 10 + 5 + \frac{1}{2} \times \pi \times 10$
$\quad = 35.7 \text{ cm}$

3 $A = \pi \times r^2$
$40 = \pi \times r^2$
$\dfrac{40}{\pi} = r^2$
$r = \sqrt{\dfrac{40}{\pi}}$
$r = 3.57 \text{ cm}$

Now try this

1 Find the area of this shape.

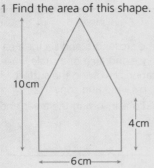

2 A semi-circular rug has a radius of 0.7 metres. Calculate the area of the rug.

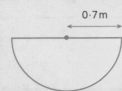

3 A circular path has an internal radius of 8 metres and an external radius of 12 metres. Find the area of the path.

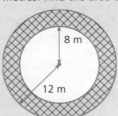

● Arc length and area of a sector of a circle

- Arc length $= \dfrac{x}{360} \times 2\pi r$
- Area of a sector $= \dfrac{x}{360} \times \pi r^2$

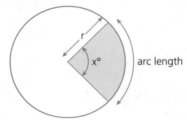

Sample question

1 a Find the arc length AB.
  b Find the shaded area.

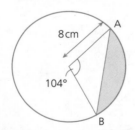

Answer

1 a Arc length $= \dfrac{104}{360} \times \pi \times 16 = 14.5$ cm

  b Segment = sector − triangle
    $= \dfrac{104}{360} \times \pi \times 8^2 - \dfrac{1}{2} \times 8 \times 8 \times \sin 104°$
    $= 27.0$ cm^2

Now try this

4 A sector of a circle of radius 5 cm has an area of 25 cm^2. Find the angle at the centre of the circle.
5 AB and CD are arcs of circles centre O. Find the perimeter ABCD of the shaded region.

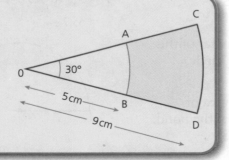

● Sample exam question and answer

A heart shape is made from a square and two semi-circles. Find the area and perimeter of the heart shape.

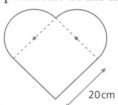

20 cm

$A = 2(\frac{1}{2} \times \pi \times 10^2) + 20 \times 20 = 714.2 \text{ cm}^2$

$P = 2(\frac{1}{2} \times \pi \times 20) + 20 + 20 = 102.8 \text{ cm}$

● Volumes and surface areas of three-dimensional shapes

Prisms

A prism is a 3-D shape with a constant cross-section. The following are examples of prisms.

For all prisms:

- Volume = area of cross-section × length (or height).
- $V = A \times L$.

The surface area of a prism is the total area of all the surfaces of the prism.

    For a cylinder of radius r and height h:

- Volume = area of cross-section × height = $\pi r^2 h$.
- Curved surface area = circumference × height = $2\pi rh$.

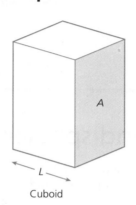

Cuboid

Cylinder

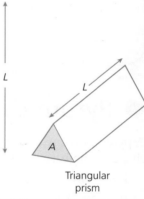

Triangular prism

Examiner's tip

Do not forget to include the units with your answer.

Sample questions

1 Find the volume and surface area of this prism.

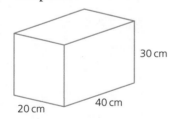

30 cm

20 cm   40 cm

2 The volume of this packet of sweets is 68 cm³. The radius is 1.2 cm.

L

Find its length.

Answers

1 Volume = area of cross-section × length
            = $(40 \times 30) \times 20$
            = $24\,000 \text{ cm}^3$

  Surface area = $(2 \times \text{top}) + (2 \times \text{side}) + (2 \times \text{front})$
                 = $2(20 \times 40) + 2(30 \times 40) + 2(20 \times 30)$
                 = 5200 cm^2

2 Volume = area of cross-section × length
      $68 = (\pi \times 1.2^2) \times L$
         $L = \dfrac{68}{(\pi \times 1.2^2)} = 15.0 \text{ cm}$

Cambridge IGCSE Mathematics Study and Revision Guide Second Edition © Brian Seager *et al.*, 2016

Now try this

6 The volume of water in this tank is 2000 cm³. Find the depth of the water.

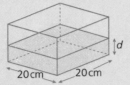

20 cm   20 cm

7 Find the volume of this bar of gold.

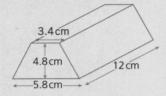

3.4 cm
4.8 cm
12 cm
5.8 cm

8 The volume of this cylinder is 50 cm³. Find the radius of the circular end.

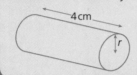

4 cm

r

● Pyramids, cones and spheres

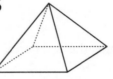

A pyramid is any 3-D shape that goes up to a point. The shape of the base is usually part of the name of the pyramid.

$V = \frac{1}{3} \times$ area of base × height.

Sample question

A pyramid has a square base 4 metres by 4 metres and is 9 metres high. Calculate the volume of the pyramid.

Answer

$V = \frac{1}{3} \times (4 \times 4) \times 9 = 48 \ m^3.$

A pyramid shape with a circular base is a cone.

- A cone has volume $V = \frac{1}{3}\pi r^2 h$ and curved surface area $= \pi r \ell$.
- A sphere has volume $V = \frac{4}{3}\pi r^3$ and surface area $= 4\pi r^2$.

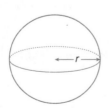

Sample question

Find the height of a cone with volume 2.5 litres and base radius 10 cm.

Answer

$2500 = \frac{1}{3} \times \pi \times 10^2 \times h$

$h = \dfrac{3 \times 2500}{\pi \times 100}$

$h = 23.9 \ cm$

It is illegal to photocopy this page

Cambridge IGCSE Mathematics Study and Revision Guide Second Edition © Brian Seager *et al.*, 2016

Sample question

A sphere has a surface area of 500 cm².
Calculate the radius of the sphere.

Answer

$$500 = 4\pi r^2$$
$$r^2 = \frac{500}{4\pi}$$
$$r = 6.3 \text{ cm}$$

Now try this

9 Find the volume of a hemisphere of radius 6 cm.
10 Find the slant height of a cone that has a base radius of 4 cm and a total surface area of 200 cm².
11 A piece of cheese is in the shape of a cuboid. One corner is cut off. The cut goes through one corner, A, and the midpoints, B and C of two sides, as shown. Calculate the volume of the remaining piece of cheese.

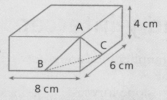

Examiner's tip

Don't forget the base in 'Now try this' question 10.

● Sample exam question and answer

Find the volume and total surface area of this prism. The ends are semi-circles.

Volume = area of cross-section × length

$$= \left(\tfrac{1}{2} \times \pi \times 2.5^2\right) \times 11 = 108 \text{ m}^3$$

Surface area $= \tfrac{1}{2} \times \pi \times 2.5^2 \times 11 + 11 \times 5 + 2 \times \tfrac{1}{2} \times \pi \times 2.5^2 = 182.6 \text{ cm}^2$

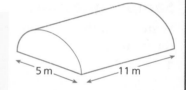

(28) Straight-line graphs

● Gradient and y-intercept

- The gradient of a line is a number indicating how steep it is. The larger the number, the steeper the line.
- Lines with positive gradient slope forwards (/); lines with negative gradient slope backwards (\).
- Gradient $= \dfrac{\text{increase in } y}{\text{increase in } x}$.
- For a line through (x_1, y_1) and (x_2, y_2), the gradient $= \dfrac{y_2 - y_1}{x_2 - x_1}$.

Sample questions

Write down the gradient and y-intercept of these lines.

1
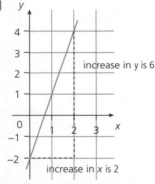

2
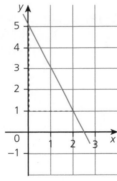

Answers

1 Gradient $= \dfrac{6}{2} = 3$; y-intercept $= -2$
   OR using $(2, 4)$ and $(0, -2)$, gradient $= \dfrac{6}{2} = 3$

2 Gradient $= \dfrac{-4}{2} = -2$; y-intercept $= 5$
   OR using $(0, 5)$ and $(2, 1)$, gradient $= \dfrac{-4}{2} = -2$

Now try this

1 Find the gradient and y-intercept of each of these lines.

a
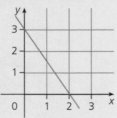

b
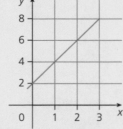

2 Find the gradient of the line that passes through each pair of coordinate points.
   a $(-4, 2)$ and $(4, 0)$ b $(0, -2)$ and $(4, 4)$ c $(-1, -3)$ and $(2, 3)$ d $(-6, 4)$ and $(6, -2)$

3 For each part of question 2, draw a diagram to find the y-intercept of the line.

Cambridge IGCSE Mathematics Study and Revision Guide Second Edition © Brian Seager *et al.*, 2016

● Line segment

A line segment is the part of a line between two points. It has finite length.
For the line segment joining (x_1, y_1) and (x_2, y_2)

$$\text{Midpoint} = \left(\frac{x_1 + x_2}{2}, \frac{y_1 + y_2}{2} \right)$$

$$\text{Length} = \sqrt{(x_2 - x_1)^2 + (y_2 - y_1)^2}$$

Sample question

A is $(1, 5)$ and B is $(7, 2)$. Find

1 a the midpoint of AB

  b the length AB

Answer

1 a $\text{Midpoint} = \left(\frac{1+7}{2}, \frac{5+2}{2} \right) = (4, 3\frac{1}{2})$

  b $\text{Length} = \sqrt{(7-1)^2 + (2-5)^2}$

$\phantom{b \text{ Length}} = \sqrt{6^2 + (-3)^2}$

$\phantom{b \text{ Length}} = \sqrt{45}$

$\phantom{b \text{ Length}} = 6.7 \text{ to 1 d.p.}$

Now try this

4 Find
  a the midpoint
  b the length joining each of these pairs of points
    (i) (4, 5) and (0, 1)   (ii) (2, 6) and (−4, 3)   (iii) (−6, 1) and (7, 5)
    (iv) (−3, −2) and (−4, −1)   (v) (2, 3) and (−1, −3)

● The general equation of a straight line $y = mx + c$

In the equation, m stands for the gradient of the line and c is the y-intercept.

The equation of a line **must** be of the form $y = mx + c$ for the two numbers to represent the gradient and y-intercept.

Sample question

Find the gradient and y-intercept of these lines.

1 $y = x + 4$

2 $2y = 6x - 3$

3 $5x - 2y = 12$

Answer

1 $y = 1x + 4$, $m = 1$, $c = 4$

2 $y = 3x - 1\frac{1}{2}$, $m = 3$, $c = -1\frac{1}{2}$

3 $-2y = -5x + 12$, so $y = 2\frac{1}{2}x - 6$, $m = 2\frac{1}{2}$, $c = -6$

Sample question

Find the equation of the line through the points $(10, 20)$ and $(30, 30)$.

Answer

$m = \frac{30-20}{30-10} = \frac{10}{20} = 0.5$, so $y = 0.5x + c$

As the line goes through $(10, 20)$, $20 = 0.5 \times 10 + c$, so $20 = 5 + c$, $c = 15$.

So the equation is $y = 0.5x + 15$.

Now try this

5 Write down the equation of each of the lines in 'Now try this' question 1.
6 Work out the gradient, m, and y-intercept, c, for each of the following straight lines
  a $y=4x-1$ c $y=x$ e $2y-3x=4$
  b $y=3+2x$ d $y+x=4$ f $x-4y=10$
7 Find the equation of the line in each part.
  a Gradient 3 and passing through the point (0, –2)
  b Gradient $-\frac{1}{2}$ and passing through the point (0, 7)
  c Gradient 2 and passing through the point (–2, 4)
  d Gradient –4 and passing through (–3, 1)
8 Find the equation of the line passing through the points
  a (1, 0) and (4, 9)   b (–2, 9) and (7, -9)

● Parallel lines

Lines that are parallel have the same gradient.

Sample question

Find the equation of the line parallel to $y=2x-5$ and passing through $(0, 4)$.

Answer

From $y=2x-5$, $m=2$

From $(0, 4)$, $c=4$

The equation is $y=2x+4$.

Now try this

9 Which of these lines are parallel?
  a $y=6x-2$ c $6y=6x-2$ e $y=6+x$
  b $y=2x-6$ d $6y=36x-24$ f $6x-y=2$

10 Find the equation of the line passing through the point (0, 7) and parallel to $y=2x$.
11 Find the equation of the line passing through the point (5, 6) and parallel to $y=3x-3$.
12 Find the equation of the line passing through the point (–3, 1) and parallel to $4y+2x=1$.
13 Find the equation of the line passing through the point (5, –1) and parallel to the line that passes through the two points (2, 3) and (5, –6).

● Perpendicular lines

Two lines with gradients m_1 and m_2 are perpendicular if $m_1 \times m_2 = -1$.

● Sample exam question and answer

Find the equation of the line perpendicular to $y=4x+1$ that passes through the point (8, 5).

$m_1=4$, so $m_2=-\frac{1}{4}$

This means that $m_2=\frac{-1}{m_1}$.

$y=-0.25x+c$.

As the line goes through $(8, 5)$, $5=-0.25 \times 8 + c$, so $5=-2+c$, $c=7$.

So the equation is $y=-0.25x+7$.

Cambridge IGCSE Mathematics Study and Revision Guide Second Edition © Brian Seager *et al.*, 2016

Now try this

14 Which of these lines are perpendicular?
    a $y=3x-5$ c $3y=5-2x$ e $2y=6x-5$
    b $2y=3x-5$ d $3y=2x-5$ f $6x-y=2$

15 Find the equation of the line passing through the point (0, 3) and perpendicular to $y=4x+5$.

16 Find the equation of the line passing through the point (–5, 2) and perpendicular to $4y+2x=1$.

17 Find the equation of the line passing through the point (6, –1) and perpendicular to the line that passes through the two points (1, –1) and (3, 5).

● Sample exam question and answer

Find the gradient and the equation of the straight line in the diagram.

Gradient $= \frac{3}{4}$

Equation is $y = \frac{3}{4}x + 2$.

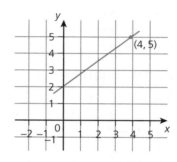

Exam-style questions

1 a Work out the gradient of this line.

   b Write down the equation of the line.

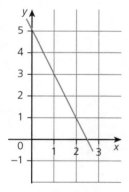

2 a Write down the gradient and *y*-intercept of the line with equation $y=4-2x$.

   b Write down the equation of the line parallel to $y=4-2x$ that passes through the point (0, –1).

3 Find the equation of the straight line that passes through the two points (–3, 12) and (5, –4).

4 Find the equation of two straight lines that pass through the point (1, 1), one parallel to $x+y=1$ and the other perpendicular to $x+y=1$.

5 a Find the equation of the line through A (6, –1) and B (3, 2).

   b Find the equation of the line perpendicular to AB that passes through A.

(29) Bearings

● Bearings

Three-figure bearings are measured clockwise from north and must have three figures. An angle of 27° would be written as a bearing of 027°.

The bearing of B from A is the angle at A, clockwise from north to the line to B, the angle marked *x* in the diagram opposite. The bearing of A from B is the angle marked *y* in the diagram.

A useful rule to remember is that the difference between the bearing of B from A and the bearing of A from B is 180°.

Often questions are asked that involve drawing distances and bearings to find places. Accurate drawing is essential in these questions.

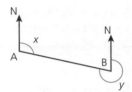

Examiner's tip

It is also useful first to do a sketch.

Sample question

Measure the bearings of A, B and C from O.

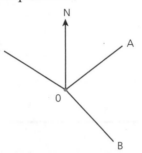

Answer

$A = 050°, B = 140°, C = 305°$

Now try this

1 Mark a point O. Draw a north line through O. Mark points A, B and C all 5 cm from O on the following bearings from O: A 060°, B 135°, C 295°.
2 The bearing of A from B is 127°. What is the bearing of B from A?
3 Anytown is 8 km from Briston and on a bearing of 125° from Briston. Make a scale drawing showing Anytown and Briston. Use a scale of 1 cm to 1 km.

● Sample exam questions and answers

Simon went orienteering. This is a sketch he made of part of the course:

1 Draw an accurate plan of this part of the course. Use a scale of 1 cm to 50 m.

2 Use your drawing to find the bearing of C from A.

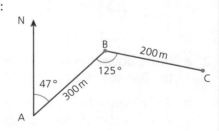

1 Diagram shown half-size.   2 069°.

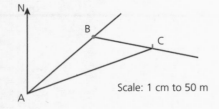

Scale: 1 cm to 50 m

Cambridge IGCSE Mathematics Study and Revision Guide Second Edition © Brian Seager *et al.*, 2016

Exam-style questions

1 A ship leaves port and travels on a bearing of 070° for 6 km to A and then on a bearing of 120° for 4 km to B.

   a Make a scale drawing of the route of the ship with a scale of 1 cm to 1 km.

   b At B, how far and on what bearing is the ship from the port?

2 P is 8 km from O on a bearing of 037° and Q is 7 km due east of O.

   a Make a scale drawing showing O, P and Q, using a scale of 1 cm to 2 km.

   b (i) Find the length of PQ.

      (ii)Find the bearing of P from Q.

3 The bearing of Corunna from Acacia is 105° and the distance is 11 km.

   a What is the bearing of Acacia from Corunna?

   b Draw a scale drawing of these places using a scale of 1 cm to 2 km.

4 The bearing of Zomba from Mousehole is 052° and the distance between them is 22 km. Liang is due west of Zomba and is on a bearing of 018° from Mousehole.

   a What is the bearing of Mousehole from

      (i) Zomba?

      (ii)Liang?

   b (i) Make a scale drawing of these places using a scale of 1 cm to 2 km.

      (ii)What is the distance between Mousehole and Liang?

Cambridge IGCSE Mathematics Study and Revision Guide Second Edition © Brian Seager *et al.*, 2016

(30) Trigonometry

Key objectives

- To apply Pythagoras' theorem and the sine, cosine and tangent ratios for acute angles to the calculation of a side or of an angle of a right-angled triangle.
- To solve trigonometrical problems in two dimensions involving angles of elevation and depression.
- To extend sine and cosine values to angles between 90° and 180°.

You should be familiar with the ratios sin, cos and tan.

$$\tan x = \frac{O}{A}$$

$$\sin x = \frac{O}{H}$$

$$\cos x = \frac{A}{H}$$

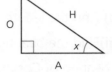

You should also be able to find the sine and cosine of any angle between 90° and 180°. $\sin(180-x) = \sin x$, $\cos(180-x) = -\cos x$.

You also need to remember Pythagoras' theorem.
$$a^2 = b^2 + c^2$$

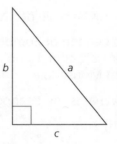

Examiner's tip

Remember to:

- label the triangle, using letters for the sides
- write down the rule
- substitute the actual values for the sides
- do any rearrangement, if necessary.

Sample questions

1 a Calculate the length of the unknown side in this triangle.

b Calculate the size of the angle marked x.

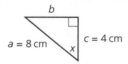

2 Find the length of AB.

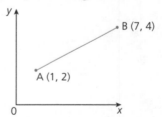

Answers

$$
\begin{aligned}
1\ a\quad a^2 &= b^2 + c^2 \\
8^2 &= b^2 + 4^2 \\
b^2 &= 64 - 16 \\
b^2 &= 48 \\
b &= \sqrt{48} \\
b &= 6.9\,\text{cm, to 1 d.p.}
\end{aligned}
$$

$$
\begin{aligned}
b\quad \cos x &= \frac{4}{8} = 0.5 \\
x &= 60°
\end{aligned}
$$

$$
\begin{aligned}
2\quad AB^2 &= AP^2 + BP^2 \\
AB^2 &= 6^2 + 2^2 = 40 \\
AB &= \sqrt{40} = 6.3
\end{aligned}
$$

Examiner's tip

Where possible, use given data rather than calculated data in later parts of the question.

Cambridge IGCSE Mathematics Study and Revision Guide Second Edition © Brian Seager *et al.*, 2016

Now try this

Find the length of the side marked *x* in each of the following. (All lengths are in cm. Give your answer correct to 1 d.p.)

1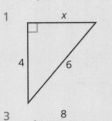

2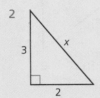

3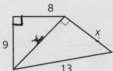

4 Find
   a AB
   b the angle ACB

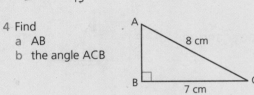

5 Find
   a DE
   b EF

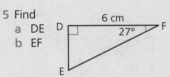

6 Find
   a MN
   b the angle LMN

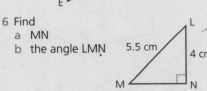

7 A ladder of length *L* is placed against a vertical wall. The ground is horizontal, the foot of the ladder is 3 m from the wall and the top is 8 m vertically above the ground.
   a Find the length of the ladder.
   b The foot of the ladder is moved a further 2 m from the base of the wall. Find the distance the ladder moves down the wall.

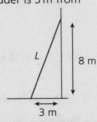

8 Find the angle PQR.

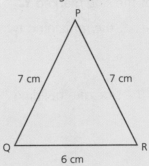

9 ABCD is the side of the frame of the goal used by a football club.
   a Calculate the length of the sloping strut, L, joining C to B.
   b Calculate the angle *x* that CB makes with the ground.

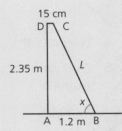

10 a Express each of the following in terms of the sine of another angle between 0° and 180°.
      (i) sin 100°, (ii) sin 153°, (iii) sin 150°, (iv) sin 60°.
   b Find all the angles between 0° and 180° that have the following sine:
      (i) 0.23, (ii) 0.47, (iii) 0.08, (iv) 0.83.

11 Express each of the following in terms of the cosine of another angle between 0° and 180°:
   (i) cos 100°, (ii) cos 153°, (iii) −cos 150°, (iv) −cos 60°.

● Sample exam question and answer

A slide in an open air swimming pool is built down the side of a bank. The slide is 8.5 m long and descends 2.6 m from the top to the water level.

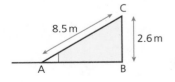

Calculate the distance AB.

8.1 m

$AC^2 = AB^2 + BC^2$

$8.5^2 = AB^2 + 2.6^2$

$AB^2 = 8.5^2 - 2.6^2$

$AB^2 = 65.49$

$AB = 8.1$

Exam-style questions

1 a Find the length of the side marked *x*.

    b Work out the sizes of the remaining two angles.

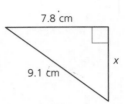

2 The drawing shows the side of a building. Calculate the length of the sloping roof.

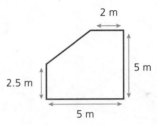

3 a Calculate the length of this ladder.

    b Work out the size of the angle between the foot of the ladder and the ground.

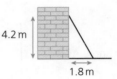

4 X, Y, Z and T are four corners of a cube of 10-cm sides. One corner is sliced off, as shown by the shaded portion, ABC. XA = 2 cm, BZ = 5 cm, CT = 3 cm.

   Calculate the lengths of

   a AB

   b AC

   c BC.

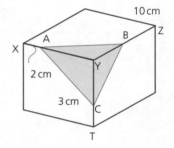

5 Tony stands at point A and flies a kite on a 40-m-long string with an angle of elevation of 36°. Colin stands at point C, directly facing Tony, and finds the angle of elevation of the kite to be 41°.

   Work out the distance between Tony and Colin.

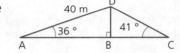

6 a On a single diagram, draw the graphs of *y* = sin θ and *y* = cos θ for 0° ≤ θ ≤ 180°.

    b From your graph, find all the angles for which sin θ and cos θ are the same.

Key objectives

- To solve problems using the sine and cosine rules for any triangle and the formula area of triangle $= \frac{1}{2}ab \sin C$.
- To solve simple trigonometrical problems in three dimensions including the angle between a line and a plane.

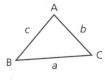

In this triangle the sine rule is $\dfrac{a}{\sin A} = \dfrac{b}{\sin B} = \dfrac{c}{\sin C}$.

The cosine rule is:

$a^2 = b^2 + c^2 - 2bc \cos A$,
$b^2 = c^2 + a^2 - 2ac \cos B$,
$c^2 = a^2 + b^2 - 2ab \cos C$.

The area of the triangle is given by $\frac{1}{2}ab \sin C$.

Sample questions

1 Find the length of AC.

2 Calculate the area of triangle ABC.

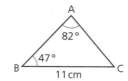

Answers

1 $\dfrac{a}{\sin A} = \dfrac{b}{\sin B}$

$\dfrac{11}{\sin 82°} = \dfrac{b}{\sin 47°}$

$b = \dfrac{11 \times \sin 47°}{\sin 82°} = 8.12\,cm$

AC = 8.12 cm

2 Angle ACB = 180 − 47 − 82 = 51

Using $\frac{1}{2}ab \sin C$,

Area $= \frac{1}{2} \times 11 \times 8.12 \times \sin 51° = 34.7\,cm^2$

Now try this

1 Find
   a AB
   b the angle ABC

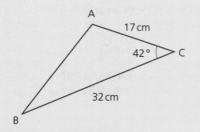

2 In triangle XYZ, angle YXZ = 130°, angle XZY = 28° and YZ = 14 cm. Calculate the length of XZ.

3 A submarine is on a bearing of 275° from a ship, Q, and on a bearing of 312° from a second ship, P. Ship P is 200 m due south of ship Q. Calculate the distance from the submarine to the second ship, P.

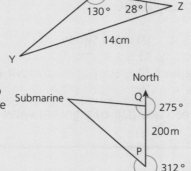

● Sample exam questions and answers

A ship sails from a port P a distance of 7 km on a bearing of 310° and then a further 11 km on a bearing of 070° to arrive at a point X where it anchors.

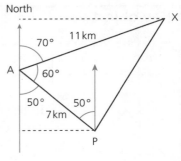

1 Calculate the distance from P to X.

2 Calculate how far east of P the point X is.

1 9.64 km
$PX^2 = 7^2 + 11^2 - 2 \times 7 \times 11 \cos 60 = 49 + 121 - 77$
$PX = 9.64$

2 4.97 km
Distance east = 11 sin 70 − 7 sin 50
         = 4.97

Exam-style questions

1 To assist hang gliders, a large orange arrow is placed on the ground to show the wind direction. The triangles ABD and CBD are congruent. BD = 10 m, DC = 8 m, angle BDC = 160°.

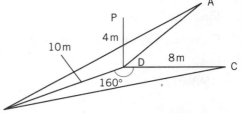

   a Calculate the length of BC.

   A vertical pole, DP, of height 4 m with a flag at the top, is fixed at D and held by wires from A, B and C.

   b Calculate the length of wire from A to P.

2 In triangle ACB, AC = 3 m, BC = 8 m and angle ACB = 15°.

   a Calculate the length of AB.

   b Calculate the area of triangle ABC.

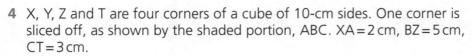

3 The diagram shows the position of three ships: A, B and C.

   a Calculate the distance BC.

   b An air–sea rescue search has to be made of the region inside triangle ABC. Calculate the area to be searched.

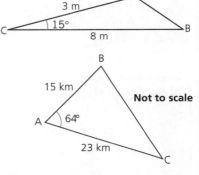

Not to scale

4 X, Y, Z and T are four corners of a cube of 10-cm sides. One corner is sliced off, as shown by the shaded portion, ABC. XA = 2 cm, BZ = 5 cm, CT = 3 cm.

   a Calculate the size of the angle ACB.

   b Calculate the area of triangle ABC.

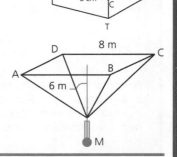

5 A, B, C and D are four points making a square of 8-m sides on the ceiling of a lecture hall at a university. M is a microphone hanging 6 m directly below the centre of the square. It is supported by four wires, one from each corner of the square.

   Calculate the angle between any pair of adjacent wires at the point where the four wires meet.

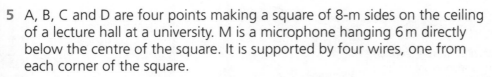

Cambridge IGCSE Mathematics Study and Revision Guide Second Edition © Brian Seager *et al.*, 2016

Key objectives

- To describe a translation by using a vector represented by, for example, $\begin{pmatrix} x \\ y \end{pmatrix}$, $\overrightarrow{AB}$ or **a**.
- To add and subtract vectors.
- To multiply a vector by a scalar.
- To calculate the magnitude of a vector $\begin{pmatrix} x \\ y \end{pmatrix}$ as $\sqrt{x^2 + y^2}$.
- To use position vectors.
- To represent vectors by directed line segments.
- To use the sum and difference of two vectors to express given vectors in terms of two coplanar vectors.

Vectors can be used to describe translations; $\overrightarrow{AB} = \begin{pmatrix} 2 \\ 3 \end{pmatrix}$ describes a translation by 2 units in the x direction and 3 units in the y direction.

Vectors are added by starting one vector where the previous one finishes.

$$\mathbf{a} = \begin{pmatrix} 3 \\ 4 \end{pmatrix}, \mathbf{b} = \begin{pmatrix} 2 \\ -1 \end{pmatrix}, \qquad \mathbf{a} + \mathbf{b} = \begin{pmatrix} 5 \\ 3 \end{pmatrix}$$

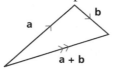

If they are in component form, vectors may be added by adding their corresponding components.

To subtract two vectors, use $\mathbf{a} - \mathbf{b} = \mathbf{a} + -\mathbf{b}$.

Vectors can be multiplied by scalars; when $\mathbf{a} = \begin{pmatrix} 2 \\ 3 \end{pmatrix}$, $4\mathbf{a} = \begin{pmatrix} 8 \\ 12 \end{pmatrix}$.

The magnitude of vector $\begin{pmatrix} x \\ y \end{pmatrix}$ is $\sqrt{x^2 + y^2}$.

$\mathbf{a} = k\mathbf{b}$ means that **a** is parallel to **b** and it is k times as long.

> **Examiner's tip**
>
> When **a** is a vector, it is displayed in bold. When you write it, you should underline the letter, a̲.

Now try this

1 a Write each of these translations as column vectors
   (i) $\overrightarrow{AD}$ (ii) $\overrightarrow{CB}$ (iii) $\overrightarrow{AB}$
   (iv) **c** (v) **a** (vi) $-\mathbf{b}$.

   b Write the following as column vectors
   (i) $\mathbf{a} + \mathbf{c}$ (ii) $\mathbf{b} - \mathbf{a}$ (iii) $\mathbf{b} - \mathbf{c}$.

   c Write the following as column vectors
   (i) $3\mathbf{a}$ (ii) $2\mathbf{b}$ (iii) $-4\mathbf{c}$.

2 Given $\mathbf{a} = \begin{pmatrix} 5 \\ 2 \end{pmatrix}$ and $\mathbf{b} = \begin{pmatrix} -1 \\ 2 \end{pmatrix}$

   a find
   (i) $\mathbf{a} + \mathbf{b}$ (ii) $\mathbf{a} - \mathbf{b}$.

   b write the following as single column vectors
   (i) $3\mathbf{a} + 2\mathbf{b}$ (ii) $2\mathbf{a} - 3\mathbf{b}$.

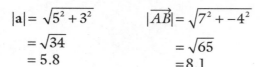

● Vector magnitude

The magnitude of a vector **a** or $\overrightarrow{AB}$ is its length and is written as $|\mathbf{a}|$ or $|\overrightarrow{AB}|$.

To find the magnitude of a vector, use Pythagoras' theorem.

$$|\mathbf{a}| = \sqrt{5^2 + 3^2}$$
$$= \sqrt{34}$$
$$= 5.8$$

$$|\overrightarrow{AB}| = \sqrt{7^2 + -4^2}$$
$$= \sqrt{65}$$
$$= 8.1$$

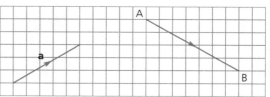

Now try this

3 Calculate the magnitude of each of these vectors

a $\mathbf{x} = \begin{pmatrix} 2 \\ 5 \end{pmatrix}$ 　　　 b $\mathbf{y} = \begin{pmatrix} 3 \\ 4 \end{pmatrix}$ 　　　 c $\mathbf{z} = \begin{pmatrix} 8 \\ -3 \end{pmatrix}$.

4 Calculate the magnitude of each of these vectors

a $\overrightarrow{AB} = \begin{pmatrix} 3 \\ 7 \end{pmatrix}$ 　　 b $\overrightarrow{PQ} = \begin{pmatrix} 6 \\ -5 \end{pmatrix}$ 　　 c $\overrightarrow{XY} = \begin{pmatrix} 3 \\ 2 \end{pmatrix}$.

5 Given the vectors $\mathbf{f} = \begin{pmatrix} -3 \\ 4 \end{pmatrix}$, $\mathbf{g} = \begin{pmatrix} 3 \\ -2 \end{pmatrix}$ and $\mathbf{h} = \begin{pmatrix} 5 \\ -3 \end{pmatrix}$, calculate the magnitude of

a $\mathbf{f}+\mathbf{h}$ 　　　　 b $2\mathbf{g}-\mathbf{f}$ 　　　　 c $\mathbf{g}+\mathbf{h}$
d $2\mathbf{g}+3\mathbf{f}$ 　　　 e $3\mathbf{h}-2\mathbf{f}$.

● Position vectors

A position vector is a vector in relation to a fixed point.

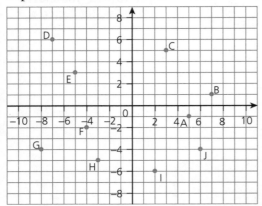

$\overrightarrow{OA} = \begin{pmatrix} 5 \\ -1 \end{pmatrix}$, $\overrightarrow{OB} = \begin{pmatrix} 7 \\ 1 \end{pmatrix}$, $\overrightarrow{OC} = \begin{pmatrix} 3 \\ 5 \end{pmatrix}$, $\overrightarrow{OD} = \begin{pmatrix} -7 \\ 6 \end{pmatrix}$, $\overrightarrow{OE} = \begin{pmatrix} -5 \\ 3 \end{pmatrix}$, $\overrightarrow{OF} = \begin{pmatrix} -4 \\ -2 \end{pmatrix}$,

$\overrightarrow{OG} = \begin{pmatrix} -8 \\ -4 \end{pmatrix}$, $\overrightarrow{OH} = \begin{pmatrix} -3 \\ -5 \end{pmatrix}$, $\overrightarrow{OI} = \begin{pmatrix} 2 \\ -6 \end{pmatrix}$, $\overrightarrow{OJ} = \begin{pmatrix} 6 \\ -4 \end{pmatrix}$.

Sample questions

Using the diagram above, write down the position vectors of the following

1 A relative to E 　　　 **2** B relative to H 　　　 **3** C relative to G

4 D relative to A 　　　 **5** H relative to E 　　　 **6** E relative to C.

Answers

1 $\begin{pmatrix} 10 \\ -4 \end{pmatrix}$ 　　　 2 $\begin{pmatrix} 10 \\ 6 \end{pmatrix}$ 　　　 3 $\begin{pmatrix} 11 \\ 9 \end{pmatrix}$

4 $\begin{pmatrix} -12 \\ 7 \end{pmatrix}$ 　　　 5 $\begin{pmatrix} 2 \\ -8 \end{pmatrix}$ 　　　 6 $\begin{pmatrix} -8 \\ -2 \end{pmatrix}$.

Cambridge IGCSE Mathematics Study and Revision Guide Second Edition © Brian Seager *et al.*, 2016

Now try this

6 On a suitable grid, plot

a A, where A has the position
  vector $\begin{pmatrix} 7 \\ 4 \end{pmatrix}$ relative to the origin

b B, where B has the position
  vector $\begin{pmatrix} -5 \\ 1 \end{pmatrix}$ relative to A

c C, where C has the position
  vector $\begin{pmatrix} -4 \\ -3 \end{pmatrix}$ relative to B

d D, where D has the position
  vector $\begin{pmatrix} 2 \\ -5 \end{pmatrix}$ relative to C

e E, where E has the position
  vector $\begin{pmatrix} -4 \\ -1 \end{pmatrix}$ relative to D

f F, where F has the position
  vector $\begin{pmatrix} 8 \\ 5 \end{pmatrix}$ relative to E.

7 a Write down the position vectors, relative to O, of points A to J.
  b Write down the position vectors of
    (i) A relative to H
    (ii) B relative to I
    (iii) C relative to G
    (iv) D relative to H
    (v) E relative to A
    (vi) F relative to J
    (vii) G relative to B
    (viii) H relative to C
    (ix) I relative to D
    (x) J relative to G.

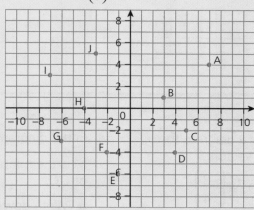

8 Show that the line joining A (–1, –3) and B (5, 6) is parallel to the vector $\begin{pmatrix} 2 \\ 3 \end{pmatrix}$.

9 The position vectors of A, B and C from O are **a**, **b** and 4**b**−3**a**, respectively.
  a Find $\overrightarrow{AB}$ in terms of **a** and **b**.
  b Show clearly that A, B and C lie in a straight line.

● Sample exam questions and answers

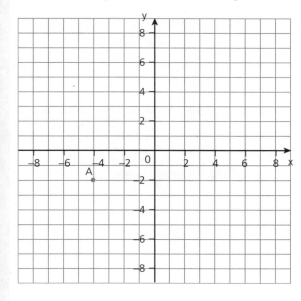

Point A is marked on the grid.

$$\overrightarrow{AB} = \begin{pmatrix} 6 \\ -3 \end{pmatrix} \text{ and } \overrightarrow{AC} = \begin{pmatrix} 9 \\ 6 \end{pmatrix}.$$

1 Copy the grid and draw $\overrightarrow{AB}$ and $\overrightarrow{AC}$ on it.

2 Write down the coordinates of B.

3 Write $\overrightarrow{BC}$ as a column vector.

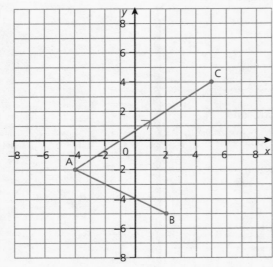

2 $B = (2, -5)$

3 $\overrightarrow{BC} = \begin{pmatrix} 3 \\ 9 \end{pmatrix}$.

Exam-style questions

1 a On a suitable grid, draw a diagram to represent the following vectors

   (i) $\overrightarrow{OX} = \begin{pmatrix} 7 \\ -2 \end{pmatrix}$ (ii) $\overrightarrow{OY} = \begin{pmatrix} -3 \\ 4 \end{pmatrix}$

   b Write $\overrightarrow{YX}$ as a column vector.

   c O, X and Y are three corners of a parallelogram.

   (i) Write down the coordinates of the fourth corner, Z.

   (ii) Write $\overrightarrow{ZO}$ as a column vector.

2 Calculate the length of $\overrightarrow{BC}$, showing your working.

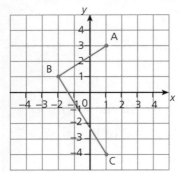

3 A and C have coordinates (–2, 1) and (2, –2), respectively.

   $AB = \begin{pmatrix} 6 \\ 5 \end{pmatrix}$. M is the midpoint of $\overrightarrow{AB}$ and N is the midpoint of AC.

   a Write down the coordinates of B, M and N.

   b Calculate the vectors $\overrightarrow{BC}$ and $\overrightarrow{MN}$.

   c State the relationship between BC and MN.

4 In the diagram $\overrightarrow{OA} = \mathbf{a}$, and $\overrightarrow{OB} = \mathbf{b}$.

   M is the midpoint of AC and B is the midpoint of OC.

   a Write down the following vectors in terms of **a** and **b**, simplifying your answers where possible

   (i) $\overrightarrow{OC}$ (ii) $\overrightarrow{BA}$ (iii) $\overrightarrow{OM}$

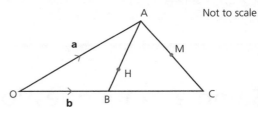

   b H is a point on $\overrightarrow{BA}$ and $\overrightarrow{BH} = \frac{1}{3}\overrightarrow{BA}$. Find the vector $\overrightarrow{OH}$ in terms of **a** and **b**.

   c What do your results tell you about the points O, H and M?

5 Draw axes from –8 to 8 for x and –4 to 4 for y.

   a Plot the points A (–7, –1), B (2, 2), C (7, 1) and D (3, –3).

   b Write down the position vector of

   (i) C relative to A

   (ii) B relative to D

   c Calculate $|\overrightarrow{AC}|$.

Cambridge IGCSE Mathematics Study and Revision Guide Second Edition © Brian Seager *et al.*, 2016

● Order

The size or **order** of a matrix is given by the number of rows and the number of columns, in that order. This is written as $m \times n$, where m is the number of rows and n is the number of columns.

Sample questions

1 Write down the order of the following matrices

a $A = \begin{pmatrix} 2 & 5 & 3 \\ 4 & 2 & 5 \end{pmatrix}$ **b** $B = \begin{pmatrix} 2 & 1 & 3 & 3 \\ 4 & 1 & 2 & 3 \\ 5 & 2 & 2 & 1 \end{pmatrix}$ **c** $C = \begin{pmatrix} 5 \\ -3 \\ 4 \end{pmatrix}$

2 A small bakery makes 45 white, 30 wholemeal and 20 granary loaves on Friday. On Saturday, it makes 40 white, 20 wholemeal and 15 granary loaves.

    Show this information in a 3×2 matrix.

Answers

1 a 2×3
  b 3×4
  c 3×1

2 $\begin{pmatrix} 45 & 40 \\ 30 & 20 \\ 20 & 15 \end{pmatrix}$

Examiner's tip

The matrix must have three rows and two columns.

Now try this

1 Write down the order of these matrices

  a $A = \begin{pmatrix} 7 & 1 & 2 & 4 \\ 2 & 5 & 3 & 8 \end{pmatrix}$ b $B = \begin{pmatrix} 5 & -3 \\ 2 & 4 \\ -1 & 5 \\ 3 & 8 \end{pmatrix}$ c $C = \begin{pmatrix} 3 & 8 & 5 & -2 \\ 1 & -3 & 2 & 5 \\ 6 & -2 & 1 & 4 \end{pmatrix}$

2 A shop sells 5 rings, 8 bracelets and 10 necklaces one week and 11 rings, 6 bracelets and 9 necklaces the next week. Show this information in a 2×3 matrix.

3 In 2001 a school had 86 students studying Spanish, 58 studying German and 102 studying English. In 2002 the numbers were 92 studying Spanish, 59 studying German and 98 studying English and in 2003 they were 95 studying Spanish, 56 studying German and 116 studying English. Show this information in a 3×3 matrix.

● Adding and subtracting

Matrices can only be added or subtracted if they are of the same order.

To add or subtract matrices, combine elements that occupy corresponding positions.

Sample questions

Work out the following

1 $\begin{pmatrix} 5 & 7 & 3 \\ 2 & 5 & 8 \end{pmatrix} + \begin{pmatrix} 3 & 2 & 6 \\ 5 & 7 & 4 \end{pmatrix}$

2 $\begin{pmatrix} 26 & 13 \\ 17 & 24 \\ 23 & 15 \end{pmatrix} - \begin{pmatrix} 18 & 11 \\ 23 & 20 \\ 19 & 23 \end{pmatrix}$

Answers

1 $\begin{pmatrix} 8 & 9 & 9 \\ 7 & 12 & 12 \end{pmatrix}$

2 $\begin{pmatrix} 8 & 2 \\ -6 & 4 \\ 4 & -8 \end{pmatrix}$

Note: $5 + 3 = 8$.

Now try this

4 Write each of the following as a single matrix

a $\begin{pmatrix} 7 & 3 & 5 & 1 \\ 4 & 5 & 3 & 6 \end{pmatrix} + \begin{pmatrix} 4 & 3 & 2 & 5 \\ 3 & 5 & 7 & 1 \end{pmatrix}$

b $\begin{pmatrix} 24 & 19 \\ -8 & 34 \\ 16 & 5 \end{pmatrix} + \begin{pmatrix} 13 & -7 \\ 14 & 1 \\ 9 & 31 \end{pmatrix}$

5 Write each of the following as a single matrix

a $\begin{pmatrix} 6 & 5 \\ 7 & 4 \end{pmatrix} - \begin{pmatrix} 4 & 7 \\ 8 & 6 \end{pmatrix}$

b $\begin{pmatrix} -2 & 5 & 8 \\ 6 & 9 & 1 \\ 3 & 7 & 2 \\ 4 & 1 & 6 \end{pmatrix} - \begin{pmatrix} 5 & 3 & 8 \\ 7 & 2 & 9 \\ 4 & 2 & 5 \\ 3 & 8 & 4 \end{pmatrix}$

● Multiplication

Matrices can be multiplied by a number, known as a *scalar*, or sometimes by another matrix.

To multiply a matrix by another matrix, the number of *columns* in the *first* matrix **must** be the same as the number of *rows* in the *second* matrix.

Sample questions

1 Work out the following

a $3\begin{pmatrix} 3 & -8 \\ 4 & 5 \end{pmatrix}$

b $\frac{1}{2}\begin{pmatrix} 4 & 9 & -2 \\ 3 & 6 & 8 \\ -4 & 1 & -3 \end{pmatrix}$

2 Where possible, multiply the following matrices

a $\begin{pmatrix} 3 & -8 \\ 4 & 5 \end{pmatrix}\begin{pmatrix} 3 & -8 \\ 4 & 5 \end{pmatrix}$

b $\begin{pmatrix} -2 & 4 & 1 \\ 3 & 0 & -5 \end{pmatrix}\begin{pmatrix} 5 & -3 \\ 2 & 1 \\ -4 & 2 \end{pmatrix}$

c $\begin{pmatrix} 4 \\ -3 \end{pmatrix}\begin{pmatrix} 3 & -8 \\ 4 & 5 \end{pmatrix}$

Answers

1 a $\begin{pmatrix} 9 & -24 \\ 12 & 15 \end{pmatrix}$ b $\begin{pmatrix} 2 & 4.5 & -1 \\ 1.5 & 3 & 4 \\ -2 & 0.5 & -1.5 \end{pmatrix}$

2 a $\begin{pmatrix} 3 \times 3 + -8 \times 4 & 3 \times -8 + -8 \times 5 \\ 4 \times 3 + 5 \times 4 & 4 \times -8 + 5 \times 5 \end{pmatrix}$

$= \begin{pmatrix} -23 & -64 \\ 32 & -7 \end{pmatrix}$

b $\begin{pmatrix} -2 \times 5 + 4 \times 2 + 1 \times -4 & -2 \times -3 + 4 \times 1 + 1 \times 2 \\ 3 \times 5 + 0 \times 2 + -5 \times -4 & 3 \times -3 + 0 \times 1 + -5 \times 2 \end{pmatrix}$

$= \begin{pmatrix} -6 & 12 \\ 35 & -19 \end{pmatrix}$

c Not possible. There is only one column in the first matrix and there are two rows in the second matrix.

Now try this

6 Where possible, multiply the following matrices

a $\begin{pmatrix} 2 & 3 & 1 \\ 4 & 5 & 3 \end{pmatrix}\begin{pmatrix} 3 & 2 \\ 2 & 4 \\ 1 & 3 \end{pmatrix}$ b $\begin{pmatrix} 3 & -2 \\ 1 & 4 \end{pmatrix}\begin{pmatrix} 5 & 2 \\ -3 & 4 \end{pmatrix}$ c $\begin{pmatrix} 4 & 2 & 5 \\ 3 & 1 & 3 \end{pmatrix}\begin{pmatrix} 3 & -2 \\ -1 & 4 \end{pmatrix}$

d $\begin{pmatrix} 6 & 3 & 2 & -1 \\ 2 & 4 & 0 & 3 \end{pmatrix}\begin{pmatrix} 3 \\ 2 \\ 2 \\ 4 \end{pmatrix}$ e $(2 \ 0 \ -3)\begin{pmatrix} 2 & 3 \\ 2 & -1 \\ 5 & 4 \end{pmatrix}$ f $(-3 \ 2)\begin{pmatrix} 0 & -3 \\ 4 & -2 \end{pmatrix}$

g $\begin{pmatrix} 1 & -2 & 3 \\ 3 & 4 & -3 \end{pmatrix}\begin{pmatrix} 2 & -3 & 4 \\ 1 & 0 & 2 \\ 3 & 5 & 2 \end{pmatrix}$ h $\begin{pmatrix} 5 & 0 & 2 \\ -3 & 1 & 4 \\ 2 & 6 & 3 \end{pmatrix}\begin{pmatrix} 2 & -3 & 4 \\ 1 & 3 & 5 \\ 0 & 4 & -2 \end{pmatrix}$

● The determinant

The determinant of a 2×2 matrix is found by calculating the difference in the products of the leading diagonal and the secondary diagonal.

For the 2×2 matrix $\begin{pmatrix} a & b \\ c & d \end{pmatrix}$, the determinant would be $\begin{vmatrix} a & b \\ c & d \end{vmatrix} = ad - bc$.

Sample question

Find the determinant |D| of the following matrix

$\begin{pmatrix} 3 & -2 \\ 1 & 4 \end{pmatrix}$

Answer

$|D| = (3 \times 4) - (1 \times -2)$

$= 12 + 2$

$= 14$

Now try this

7 Find the determinant of each of the following matrices

a $\begin{pmatrix} 5 & 2 \\ 4 & 3 \end{pmatrix}$
b $\begin{pmatrix} 3 & 4 \\ 5 & 8 \end{pmatrix}$
c $\begin{pmatrix} 8 & 5 \\ 5 & 3 \end{pmatrix}$
d $\begin{pmatrix} -3 & 3 \\ 5 & -6 \end{pmatrix}$

e $\begin{pmatrix} 4 & -3 \\ 2 & 5 \end{pmatrix}$
f $\begin{pmatrix} -3 & 2 \\ -4 & 5 \end{pmatrix}$
g $\begin{pmatrix} -4 & -3 \\ 5 & -2 \end{pmatrix}$
h $\begin{pmatrix} -6 & -5 \\ -4 & -3 \end{pmatrix}$

● The inverse of a matrix

When two matrices multiply together to give $\begin{pmatrix} 1 & 0 \\ 0 & 1 \end{pmatrix}$ (that is, the **identity** matrix) as the outcome, then one of the matrices is the inverse of the other.

There are three steps to find an inverse matrix for $\begin{pmatrix} a & b \\ c & d \end{pmatrix}$.

Step 1: Reverse the position of the numbers in the leading diagonal.
Step 2: Change the signs of the numbers in the secondary diagonal.
Step 3: Divide all the elements by the determinant.

$$\begin{pmatrix} \dfrac{d}{ad-bc} & \dfrac{-b}{ad-bc} \\ \dfrac{-c}{ad-bc} & \dfrac{a}{ad-bc} \end{pmatrix}$$

Sample questions

Work out the inverse matrix of

1 A, where $A = \begin{pmatrix} 8 & 5 \\ 3 & 2 \end{pmatrix}$

2 B, where $B = \begin{pmatrix} 5 & 3 \\ 6 & 4 \end{pmatrix}$

Examiner's tip

It does not matter whether you multiply the matrix by its inverse or vice versa.

Answers

1 $A^{-1} = \begin{pmatrix} 2 & -5 \\ -3 & 8 \end{pmatrix}$

2 $B^{-1} = \begin{pmatrix} 2 & -1.5 \\ -3 & 2.5 \end{pmatrix}$.

Examiner's tip

The inverse of matrix A is written A^{-1}.

Check your answers by multiplying

1 $\begin{pmatrix} 8 & 5 \\ 3 & 2 \end{pmatrix}\begin{pmatrix} 2 & -5 \\ -3 & 8 \end{pmatrix} = \begin{pmatrix} 1 & 0 \\ 0 & 1 \end{pmatrix}$

2 $\begin{pmatrix} 5 & 3 \\ 6 & 4 \end{pmatrix}\begin{pmatrix} 2 & -1.5 \\ -3 & 2.5 \end{pmatrix} = \begin{pmatrix} 1 & 0 \\ 0 & 1 \end{pmatrix}$

Determinant is $16 - 15 = 1$

Determinant is $20 - 18 = 2$

Now try this

8 Find the inverse matrix of each of the following matrices

a $\begin{pmatrix} 7 & 5 \\ 4 & 3 \end{pmatrix}$
b $\begin{pmatrix} 2 & 3 \\ 3 & 5 \end{pmatrix}$
c $\begin{pmatrix} 7 & 9 \\ 3 & 4 \end{pmatrix}$
d $\begin{pmatrix} -8 & 3 \\ 5 & -2 \end{pmatrix}$

e $\begin{pmatrix} 4 & 2 \\ 8 & 5 \end{pmatrix}$
f $\begin{pmatrix} 6 & 4 \\ 5 & 5 \end{pmatrix}$
g $\begin{pmatrix} -5 & 7 \\ 4 & -6 \end{pmatrix}$
h $\begin{pmatrix} 6 & 8 \\ 4 & 5 \end{pmatrix}$

Cambridge IGCSE Mathematics Study and Revision Guide Second Edition © Brian Seager *et al.*, 2016

● Sample exam questions and answers

Given the matrices A $= \begin{pmatrix} 6 & 4 \\ 2 & 1 \end{pmatrix}$ and B $= \begin{pmatrix} 3 \\ 5 \end{pmatrix}$, work out

1 AB

2 A^{-1}, the inverse of A

1 $\begin{pmatrix} 6 & 4 \\ 2 & 1 \end{pmatrix}\begin{pmatrix} 3 \\ 5 \end{pmatrix} = \begin{pmatrix} 38 \\ 11 \end{pmatrix}$ 2 $\begin{pmatrix} -0.5 & 2 \\ 1 & -3 \end{pmatrix}$

Determinant is $6 - 8 = -2$

Exam-style questions

1 Given the matrices A $= \begin{pmatrix} 3 & 4 \\ 2 & 4 \end{pmatrix}$ and B $= \begin{pmatrix} 5 \\ 1 \end{pmatrix}$, work out

   a AB   b A^{-1}, the inverse of A

2 A $= \begin{pmatrix} 3 & -2 \\ 1 & 2 \end{pmatrix}$

   a Find the 2×2 matrix M, such that A $+$ M $= \begin{pmatrix} 0 & 0 \\ 0 & 0 \end{pmatrix}$.

   b Find the 2×2 matrix N, such that AN $= \begin{pmatrix} 1 & 0 \\ 0 & 1 \end{pmatrix}$.

3 P $= \begin{pmatrix} 1 & 0 \\ x & 0 \end{pmatrix}$, Q $= \begin{pmatrix} 3 & -2 \\ -2 & 3 \end{pmatrix}$ and R $= \begin{pmatrix} 4 & -2 \\ -6 & 3 \end{pmatrix}$

   a P $+$ Q $=$ R.
      Find the value of x.

   b Explain why R cannot have an inverse matrix.

   c Find Q^{-1}, the inverse of Q.

4 X $= \begin{pmatrix} 3 & 2 \\ -4 & -1 \end{pmatrix}$, Y $= \begin{pmatrix} 5 & p \\ 2 & -4 \end{pmatrix}$ and Z $= \begin{pmatrix} 19 & q \\ -22 & 16 \end{pmatrix}$

   a If XY $=$ Z, find the value of p and the value of q.

   b Find X^{-1}, the inverse of X.

5 A $= \begin{pmatrix} 1 \\ 3 \end{pmatrix}$ and B $= \begin{pmatrix} 3 & -4 \\ -2 & 5 \end{pmatrix}$

   a Explain why it is not possible to work out AB.

   b Work out

      (i) B^2

      (ii) B^{-1}, the inverse of B

Transformations

Key objectives

- To use the following transformations of the plane: reflection (M), rotation (R), translation (T), enlargement (E) and their combinations.
- To identify and give precise descriptions of transformations connecting given figures.
- To describe transformations using coordinates and matrices (singular matrices are excluded).

● Transformations

You need to know about **reflection**, **rotation**, **translation** and **enlargement**. You may have to draw a transformation or a combination of transformations. You may have to describe transformations. Always state the type of transformation and then the details.

● Reflection

The image is the same shape and size as the object but reversed. Corresponding points are on opposite sides of the **mirror line**.

To describe a reflection, state the mirror line.

● Rotation

The image is the same shape and size as the object but turned round a **centre of rotation**.

To describe a rotation, state the centre of rotation, the angle and the direction – clockwise or anticlockwise.

● Translation

The image is the same shape and size as the object and the same way up. Each point moves the same distance and direction.

To describe a translation, state the column vector or how far the object has moved and in what direction.

● Enlargement

The image is the same shape as the object but each length is multiplied by a **scale factor**. If the scale factor is negative, the image is on the opposite side of the centre from the object.

To describe an enlargement, state the centre of enlargement and the scale factor.

Cambridge IGCSE Mathematics Study and Revision Guide Second Edition © Brian Seager *et al.*, 2016

Sample question

Enlarge the flag ABCD with centre (–3, 0) and scale factor –0.5.

Answer

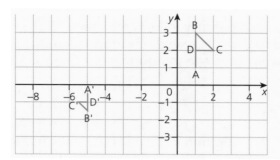

Now try this

1 Draw axes from –6 to 6 for *x* and *y*. Plot and join (3, 1), (1, 1) and (1, 4) and label this triangle A.
   a Reflect A in the *x*-axis and label the result B.
   b Reflect A in the line *x* = –1 and label the result C.

2 Draw axes from 0 to 5 for *x* and *y*. Plot the points (3, 1), (3, 2), (3, 3) and (4, 2) and join them to form flag A. Enlarge flag A with centre (1, 0) and scale factor $\frac{1}{2}$. Label the image B.

3 In this diagram, describe the transformation which maps
   a E onto A
   b A onto D
   c A onto B
   d B onto A
   e F onto G
   f B onto H
   g D onto E
   h C onto F
   i E onto D
   j C onto E
   k A onto H

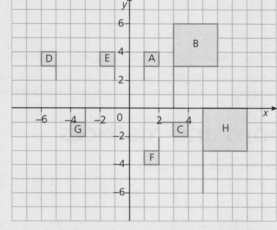

4 Draw axes from –6 to 6 for *x* and *y*. Plot and join (3, 1), (1, 1) and (1, 4) and label this triangle A.
   a Rotate A through 180° about (0, 0) and label the result B.
   b Rotate A through 90° anticlockwise about (0, 1) and label the result C.

5 Find the centre and scale factor of the following enlargement:

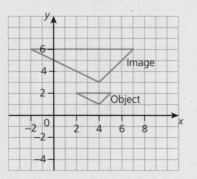

6 Draw axes from 0 to 5 for *x* and *y*. Plot the points (3, 1), (3, 2), (3, 3) and (4, 2) and join them to form flag A. Translate flag A by $\begin{pmatrix} 2 \\ -1 \end{pmatrix}$. Label the image B.

● Combining transformations

Two or more transformations may follow each other.

You may be asked to describe a single transformation that would have the same effect as two transformations that you have done.

Examiner's tip

Take care that you label diagrams carefully and transform the shape you have been asked to and not another one!

Sample question

B is the image of A after reflection in the y-axis. C is the image of B after reflection in the x-axis. Describe the single transformation that maps A directly onto C.

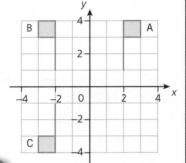

Answer

A has been rotated through 180° with centre (0, 0).

Now try this

In the following questions, carry out the transformations on a simple shape of your choice, then describe the single transformation that is equivalent to the combination.

7 A translation of $\begin{pmatrix} 1 \\ 4 \end{pmatrix}$ followed by a translation of $\begin{pmatrix} 2 \\ -3 \end{pmatrix}$.

8 A reflection in the line $x = 3$ followed by a reflection in the line $y = 2$.

9 A reflection in the x-axis followed by a rotation through 90° anticlockwise about (0, 0).

10 A rotation through 90° anticlockwise about (0, 0) followed by a reflection in the x-axis.

● Transformations and matrices

The coordinates of a shape can be given by a matrix.

For example, the quadrilateral A (1, 1), B (3, 1), C (3, 2), D (1, 3) can be defined by the following matrix: $\begin{pmatrix} 1 & 3 & 3 & 1 \\ 1 & 1 & 2 & 3 \end{pmatrix}$.

A transformation can also be given by a matrix, for example $\begin{pmatrix} -1 & 0 \\ 0 & 1 \end{pmatrix}$.

Multiplying the two matrices gives the coordinates of the image:

$$\begin{pmatrix} -1 & 0 \\ 0 & 1 \end{pmatrix}\begin{pmatrix} 1 & 3 & 3 & 1 \\ 1 & 1 & 2 & 3 \end{pmatrix} = \begin{pmatrix} -1 & -3 & -3 & -1 \\ 1 & 1 & 2 & 3 \end{pmatrix}.$$

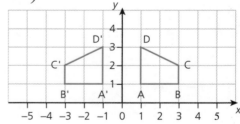

Sample questions

Triangle PQR with coordinates P (1, 1), Q (4, 2) and R (2, 4) is mapped onto P′Q′R′ using the matrix A $\begin{pmatrix} 2 & 1 \\ 1 & 2 \end{pmatrix}$.

Triangle P′Q′R′ is then transformed onto P″Q″R″ using matrix B, $\begin{pmatrix} -1 & 0 \\ 0 & 1 \end{pmatrix}$.

1 What single matrix would map triangle PQR onto triangle P″Q″R″?

2 What are the coordinates of triangle P″Q″R″?

Answers

1 The single transformation has the matrix $BA = \begin{pmatrix} -1 & 0 \\ 0 & 1 \end{pmatrix}\begin{pmatrix} 2 & 1 \\ 1 & 2 \end{pmatrix} = \begin{pmatrix} -2 & -1 \\ 1 & 2 \end{pmatrix}.$

2 $\begin{pmatrix} -2 & -1 \\ 1 & 2 \end{pmatrix}\begin{pmatrix} 1 & 4 & 2 \\ 1 & 2 & 4 \end{pmatrix} = \begin{pmatrix} -3 & -10 & -8 \\ 3 & 8 & 10 \end{pmatrix}$

The coordinates are P''(−3, 3), Q'' (−10, 8) and R''(−8, 10).
Check:

$\begin{pmatrix} 2 & 1 \\ 1 & 2 \end{pmatrix}\begin{pmatrix} 1 & 4 & 2 \\ 1 & 2 & 4 \end{pmatrix} = \begin{pmatrix} 3 & 10 & 8 \\ 3 & 8 & 10 \end{pmatrix}$

$\begin{pmatrix} -1 & 0 \\ 0 & 1 \end{pmatrix}\begin{pmatrix} 3 & 10 & 8 \\ 3 & 8 & 10 \end{pmatrix} = \begin{pmatrix} -3 & -10 & -8 \\ 3 & 8 & 10 \end{pmatrix}$

The following table gives the matrices for some common transformations:

Matrix	Transformation
$\begin{pmatrix} 1 & 0 \\ 0 & -1 \end{pmatrix}$	Reflection in the x-axis
$\begin{pmatrix} 0 & 1 \\ 1 & 0 \end{pmatrix}$	Reflection in the line $y = x$
$\begin{pmatrix} -1 & 0 \\ 0 & -1 \end{pmatrix}$	Rotation of 180° about the origin
$\begin{pmatrix} p & 0 \\ 0 & p \end{pmatrix}$	Enlargement of scale factor p and centre (0, 0)
$\begin{pmatrix} -1 & 0 \\ 0 & 1 \end{pmatrix}$	Reflection in the y-axis
$\begin{pmatrix} 0 & -1 \\ -1 & 0 \end{pmatrix}$	Reflection in the line $y = -x$
$\begin{pmatrix} 0 & -1 \\ 1 & 0 \end{pmatrix}$	Rotation of 90° anticlockwise about the origin

If you are unsure what effect multiplying by a matrix will have, then sketch the unit square and multiply its matrix $\begin{pmatrix} 0 & 0 & 1 & 1 \\ 0 & 1 & 1 & 0 \end{pmatrix}$ by the matrix you wish to check.

Examiner's tip

Remember, in matrix multiplication, the order matters.

It is illegal to photocopy this page

119

Now try this

11   A square ABCD has vertices at A (2, 0), B (4, 0), C (4, 2) and D (2, 2).

The matrix P = $\begin{pmatrix} 1 & 0 \\ 0 & -1 \end{pmatrix}$ and the matrix Q = $\begin{pmatrix} 0 & 1 \\ -1 & 0 \end{pmatrix}$.

   a   Give the coordinates of A′B′C′D′ when ABCD is transformed using matrix P.
   b   Give the coordinates of A″B″C″D″ when ABCD is transformed using matrix Q.
   c   Give the coordinates of A‴B‴C‴D‴ when A′B′C′D′ is transformed using matrix Q.
   d   Find the matrices R and S, where R = QP and S = PQ.
   e   Find the coordinates of the image of ABCD under the transformation given by R.
   f   Find the coordinates of ABCD under the transformation given by S.

12   The vertices of a triangle P are (2, 1), (4, 1) and (4, 4). The matrix M = $\begin{pmatrix} -1 & 0 \\ 0 & -1 \end{pmatrix}$ and the

matrix N = $\begin{pmatrix} -1 & 0 \\ 0 & 1 \end{pmatrix}$.

   a   (i)   Find the coordinates of the image of P under the transformation M, i.e. M(P).
       (ii) Describe the transformation produced by this matrix.
   b   (i)   Find the coordinates of the image of P under the transformation NM, i.e. NM(P).
       (ii) Describe the single transformation produced by these matrices.

● Sample exam questions and answers

Triangle T is rotated 90° clockwise about the point (0, 0). Its image is triangle S. Triangle S is reflected in the y-axis. Its image is triangle R.

1   Draw triangle S and triangle R on the diagram. Label each triangle clearly.

2   Describe a single transformation that maps R onto T.

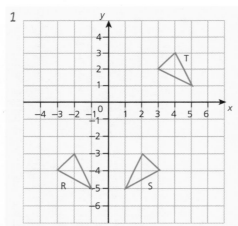

2   A reflection in the line y = −x.

Cambridge IGCSE Mathematics Study and Revision Guide Second Edition   © Brian Seager *et al.*, 2016

Exam-style questions

1 The diagram shows shapes A and B:

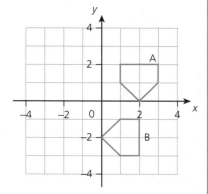

  a Describe fully the single transformation that maps shape A onto shape B.

  b Draw the image of shape A after a reflection in the *y*-axis. Label this shape C.

  c Draw the image of shape A after an enlargement. Use (0, 0) as the centre and scale factor 3. Label this shape D.
  Note that you will need an *x*-axis from −5 to 10 and a *y*-axis from −5 to 8.

2 Describe the enlargement that is equivalent to a rotation through 180° about (0, 0).

3 The diagram shows triangles T_1, T_2 and T_3.

  a T_2 is a reflection of T_1 in the *y*-axis.

   Find the matrix P representing the transformation that maps T_1 onto T_2.

  b T_3 is the image of T_2 under a 90° rotation about the origin. Find the matrix Q associated with this transformation.

  c State the single transformation that is defined by the matrix QP.

  d Explain why the single transformation defined by the matrix PQ is not the inverse of the single matrix defined by QP.

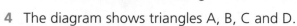

4 The diagram shows triangles A, B, C and D.

  a Describe fully the single transformation that maps triangle A onto triangle B.

  b Find the matrix representing the transformation that maps triangle A onto triangle C.

  c Triangle A is mapped onto triangle D by an enlargement. Find

   (i) the scale factor of this enlargement

   (ii) the coordinates of the centre of this enlargement.

  d The matrix $\begin{pmatrix} 2 & 1 \\ 0 & 3 \end{pmatrix}$ represents the transformation that

   maps triangle A onto triangle E. Find

   (i) the coordinates of the vertices of triangle E

   (ii) the area of triangle E

   (iii) the matrix representing the transformation that maps triangle E onto triangle A.

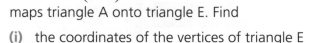

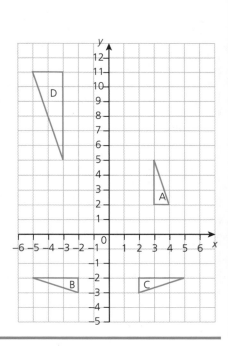

Key objectives

- To calculate the probability of a single event as a fraction, a decimal or a percentage.
- To understand and use the probability scale from 0 to 1.
- To understand that the probability of an event occurring = 1 – the probability of the event not occurring.
- To understand relative frequency as an estimate of probability.

● Basic probability

Probability can be expressed as fractions, decimals or percentages. All probabilities are between 0 and 1, inclusive.

P(A does not happen) = 1 – P(A)

For equally likely outcomes:

$$P(A) = \frac{\text{number of ways } A \text{ can happen}}{\text{total possible number of outcomes}}$$

Mutually exclusive outcomes are those that cannot happen together. If A, B and C are mutually exclusive outcomes covering all the possibilities, then P(A) + P(B) + P(C) = 1.

Examiner's tip

P(A) is a useful shorthand for the probability that A happens.

Examiner's tip

Never write probabilities as '1 in 5' or as a ratio '1 to 5' or '1:5'. You will lose marks if you do. Instead, write $\frac{1}{5}$ or 0.2 (or 20% if the question uses percentages).

Sample question

A bag contains five red balls, three green balls, four white balls and no others. One ball is drawn out at random. Work out the following probabilities:

a P(red)

b P(green)

c P(white)

d P(black)

Answer

There are 12 balls in total.

a $\frac{5}{12}$ b $\frac{3}{12} = \frac{1}{4}$ c $\frac{4}{12} = \frac{1}{3}$ d 0.

Now try this

1 Coloured sweets are packed in bags of 20. There are five different colours. The probabilities for four are given in the following table:

Colour	Orange	White	Yellow	Green	Red
Probability	0.05		0.2	0.25	0.35

a Find the probability of picking a white sweet.
b Find the probability of not picking a green sweet.
c How many sweets of each colour would you expect to find in a bag?

2 A manufacturer makes flags with three stripes. The flags have three stripes in each of the following colours: amber (A), blue (B) and cream (C). The stripes can be in any order.
a On a copy of this table, list all of the orders in which the colours can appear. The first one has been done for you.

Top	Middle	Bottom
A	B	C

b One of each of the different flags is stored in a box. Naveed takes one out at random. What is the probability that its middle colour is blue?

Cambridge IGCSE Mathematics Study and Revision Guide Second Edition © Brian Seager *et al.*, 2016

● Relative frequency

$$\text{Relative frequency} = \frac{\text{number of times event occurs}}{\text{total number of trials}}$$

When theoretical probabilities are not known, relative frequency can be used to estimate probability.

The greater the number of trials, the better the estimate. The graph of relative frequency against the number of trials may vary greatly at first, but later settles down.

Relative frequency experiments may be used to test for bias, e.g. to see if a dice is fair.

Sample question

Sarah experiments with a biased dice. This graph shows the relative frequency of throwing a six, when Sarah threw her dice 1000 times. What is the probability of getting a six with this dice?

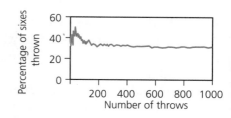

Answer

P(6) = 30% approximately.

Now try this

3 These are the results of 500 spins with a spinner:

Number on spinner	1	2	3	4	5
Frequency	106	92	74	127	101

Find as a decimal the relative frequency of scoring three with this spinner. Do these results suggest the spinner is biased? Explain your answer.

4 250 people were asked their favourite colour from a packet of sweets. 24 said mauve, 52 said yellow, 15 said brown, 72 said red and 44 said green. The other colour was orange.
 a How many said orange?
 b Estimate the probability that the next person asked says red.

● Sample exam questions and answers

A compact disc player selects at random from the tracks on the disc to be played.

1 A disc has nine tracks on it. The tracks are numbered 1, 2, 3, 4, 5, 6, 7, 8, 9. What is the probability that the first track played is

a 5 **b** 10 **c** a multiple of 4?

2 Another disc has three tracks on it. The tracks are numbered 1, 2 and 3.

a List all the different orders in which the tracks can be played. Two have been done for you:

1, 2, 3
1, 3, 2

b What is the probability that the tracks are *not* played in the order 1, 2, 3?

1 a $\frac{1}{9}$ b 0 c $\frac{2}{9}$

2 a 1, 2, 3 1, 3, 2 2, 1, 3 2, 3, 1 3, 2, 1 3, 1, 2. b $\frac{5}{6}$

Examiner's tip

Be systematic so that you do not miss out any possibilities.

It is illegal to photocopy this page

Key objectives

- To calculate the probability of simple combined events, using possibility diagrams and tree diagrams, where appropriate.

● Combining probabilities

Equally likely outcomes may be listed in a table or shown on a grid.

When outcomes are not equally likely, use **tree diagrams**. Each set of branches shows the possible outcomes of the event and probabilities, which add up to 1.

When A and B are **independent** events, the outcome of B is not affected by the outcome of A and P(A and B) = P(A) × P(B).

If A and B are **mutually exclusive**, P(A or B) = P(A) + P(B).

Sample question

This tree diagram shows the probabilities that Parinda has to stop at lights or a crossing on her way to work:

What is the probability that she does not stop at either?

Lights   **Crossing**

0.4 — stop
  0.2 — stop
  0.8 — not stop

0.6 — not stop
  0.2 — stop
  0.8 — not stop

Answer

P(not stop at either) = 0.6 × 0.8 = 0.48.

Now try this

1 Juan offers her guests chicken fried rice or prawn chow mein for main course, and blackcurrant cheesecake, ice cream or banana fritter for dessert. List the possible two-course meals her guests can have.

2 Draw a grid, with axes marked from 1 to 4, to show the possible outcomes when a fair spinner numbered 1 to 4 is spun twice. What is the probability of
a getting 1 both times
b getting a total of 6?

3 A bakery makes doughnuts. Machine A fills them with jam. There is a probability of 0.03 that machine A misses a doughnut. Machine B covers them with sugar. The probability that machine B misses a doughnut is 0.02.
a Draw a tree diagram to show these probabilities, using the first set of branches for machine A.
b Calculate the probability that a doughnut has no jam inside and no sugar.
c Calculate the probability that a doughnut has just one of these faults.

4 Bashir experiments with a biased dice. The probability that it shows a six is $\frac{2}{5}$. Draw a tree diagram and find the probability that, when the dice is thrown three times, there will be
a three sixes
b no sixes
c just one six.

● Conditional probabilities

When events are not independent, the outcome of one affects the probability that the other happens.

In a tree diagram in this case, the probabilities on the second pairs of branches will be different.

Examiner's tip

Remember that the sum of the probabilities of all the outcomes is 1.

Cambridge IGCSE Mathematics Study and Revision Guide Second Edition © Brian Seager *et al.*, 2016

Sample question

The probability that Pali wakes up late on a work morning is 0.1. When he wakes up late, the probability that he misses the bus is 0.8. When he does not wake up late, the probability that he misses the bus is 0.2. Draw a tree diagram to represent this, and find the probability that he misses the bus on a work morning.

Answer

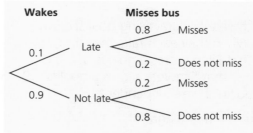

$P(\text{miss bus}) = 0.1 \times 0.8 + 0.9 \times 0.2$

$= 0.08 + 0.18 = 0.26.$

Now try this

5 A bag contains five white, two orange and three black beads. A bead is selected at random and its colour noted. It is not replaced. A second bead is then selected. Calculate the probability that the two beads are
   a   both black   b   different colours.

6 A driver has to go through two sets of traffic lights on her way to work. The probability that she has to stop at the first set of lights is $\frac{2}{3}$. If she stops at the first set of lights, the probability that she has to stop at the second set is $\frac{3}{4}$. If she does not stop at the first set of lights, the probability that she has to stop at the second set is $\frac{1}{10}$. On any particular day, what is the probability that
   a   she has to stop at just one set of lights on her way to work
   b   she has to stop *at least* once at the traffic lights on her way to work?

7 On any morning, the probability that I leave home late is $\frac{1}{3}$. If I leave late, the probability that I arrive late for work is $\frac{4}{5}$. If I leave on time, the probability that I arrive late is $\frac{2}{5}$. What is the probability that I shall be late for work tomorrow?

8 There are 15 pens in a box. Five are red and ten are green. Without looking, I take a pen from the box and do not replace it. I do this twice. What is the probability that the pens are different colours?

● Sample exam questions and answers

Pete likes crisps. Without looking, he picks a bag out of an assorted pack. There are 12 bags of crisps in the pack. Two of these bags are ready salted. The manufacturers say that one bag in every 100 has a gold reward in it. This is independent of the flavour of the crisps.

1 Complete the tree diagram to show the probabilities.

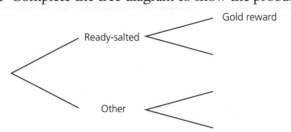

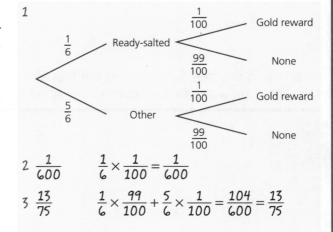

2 $\frac{1}{600}$ $\frac{1}{6} \times \frac{1}{100} = \frac{1}{600}$

3 $\frac{13}{75}$ $\frac{1}{6} \times \frac{99}{100} + \frac{5}{6} \times \frac{1}{100} = \frac{104}{600} = \frac{13}{75}$

2 What is the probability that Pete picks a ready salted bag with a gold reward in it?

3 What is the probability that Pete picks either a ready salted bag or one with a gold reward in it but not both of these?

Exam-style questions

1 Ellen and Schweta are practising goal scoring in football. When they each take one shot, the probability that Ellen scores a goal is 0.7 and the probability that Schweta scores a goal is 0.8. These probabilities are independent.

a Complete the tree diagram.

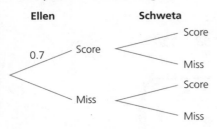

Ellen   Schweta

b Calculate the probability that, at their next attempt:

(i) both Ellen and Schweta score a goal

(ii) either Ellen or Schweta, but not both, scores a goal.

2 Carlotta and Vimal are both members of a walking club. Carlotta goes to 60% of the club's meetings. If Carlotta goes to a meeting, the probability that Vimal will go is $\frac{2}{3}$. If Carlotta does not go to a meeting, the probability that Vimal will go is $\frac{1}{4}$.

a Use this information to complete this tree diagram.

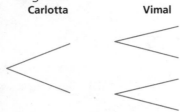

Carlotta   Vimal

b Using this tree diagram, find the probability that, for a randomly chosen meeting:

(i) both Carlotta and Vimal will go

(ii) neither of them will go.

3 James keeps his socks separately in a drawer. The drawer contains four red socks, five white socks and six black socks. He dresses in the dark one morning and pulls out two socks without being able to see their colour. What is the probability that he takes:

a two black socks

b two socks of the same colour

c two socks of different colours?

4 a Jake conducts an experiment with a fair six-sided ordinary dice. He throws it twice. What is the probability that he throws a six both times?

b He throws the dice *n* times. Write down an expression in terms of *n* for the probability that:

(i) he does not get a six on any throw

(ii) he throws at least one six.

5 Box A contains pens. 20% are red, 50% are blue and 30% are black. Box B contains pencils. 15% are blue, 50% are red and 35% are black. Jack takes a pen from Box A and a pencil from Box B at random.

a Draw a tree diagram to show the possible outcomes.

b Work out the probability that he takes a red pen and a red pencil.

c Work out the probability that he takes a pen and pencil of the same colour.

Mean, median, mode and range

Key objectives

- To calculate the mean, median, mode and range for individual and discrete data and distinguish between the purposes for which they are used.
- To calculate an estimate of the mean for grouped and continuous data.
- To identify the modal class from a grouped frequency distribution.

● Mean, median, mode and range

There are three types of 'average' that you need to know: mean, median and mode.

The **mean** is found by adding all the values and dividing the total by the number of values used.

The **median** is the middle value when all the values are arranged in order of size.

The **mode** is the value that occurs most often in the data.

The **range** measures the spread of the data and is the difference between the highest and lowest values.

Sample questions

1 For the following data:

3, 7, 4, 3, 2, 6, 8, 1, 2, 7, 3, 3, 9, 6, 5

a Find the mean, median and mode.

b Find the range.

2 The bar chart shows the scores out of 10 achieved by a group of students in a spelling test. Find the mean, median and mode for the data.

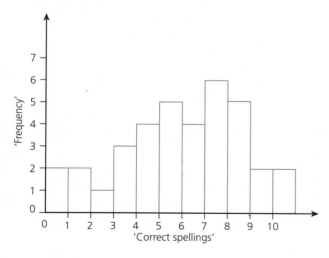

Answers

1 a $\text{Mean} = \dfrac{3+7+4+3+2+6+8+1+2+7+3+3+9+6+5}{15} = 4.6$

Median = 4 (this is the middle value of 1, 2, 2, 3, 3, 3, 3, 4, 5, 6, 6, 7, 7, 8, 9).

Mode = 3 (this appears four times).

b Range = 8 (this is 9 − 1).

2

Correct spellings	Frequency	Correct spellings × frequency
0	2	0
1	2	2
2	1	2
3	3	9
4	4	16
5	5	25
6	4	24
7	6	42
8	5	40
9	2	18
10	2	20
Total	**36**	**198**

Mean $= \frac{198}{36} = 5.5$

Median $= 6$, as there are two middle values and both are the same:

0, 0, 1, 1, 2, 3, 3, 3, 4, 4, 4, 4, 5, 5, 5, 5, 5, 6, 6, 6, 6, 7, 7, 7, 7, 7, 7, 8, 8, 8, 8, 8, 9, 9, 10, 10.

Mode $= 7$, as this appears six times.

Now try this

1 The heights, in centimetres, of the 11 members of a football team are: 190, 179, 181, 184, 186, 174, 171, 174, 184, 183, 174.
  a What is the range of the heights?   b What is the modal height?
  c What is the median height?   d Calculate the mean height of the team.

2 A quality-control process involves weighing 80 jars of jam. The weights are shown in the following table:

Weight (g)	Frequency
454	7
455	10
456	25
457	30
458	8

  a What is the modal weight?   b What is the median weight?
  c Calculate the mean weight.   d What is the range of the weights?

● Sample exam questions and answers

The labels on jars of honey show that they contain 454 g of honey. The actual weight of honey in 20 jars was checked. The following table shows the results.

Weight (g)	Number of jars (f)
454	0
455	1
456	6
457	7
458	3
459	1
460	2

1 Work out the range of the weights.

2 Calculate the mean weight of honey per jar.

Cambridge IGCSE Mathematics Study and Revision Guide Second Edition © Brian Seager *et al.*, 2016

Weight (g)	Number of jars (f)	Weight × number of jars
454	0	0
455	1	455
456	6	2736
457	7	3199
458	3	1374
459	1	459
460	2	920
Total	**20**	**9143**

1 $460 - 454 = 6$.

2 Mean $= \dfrac{9143}{20} = 457.15$ g.

● Mean from grouped data

The mean can be calculated for grouped data using mid-interval values. However, the value found is only an estimate, as not all individual values are available.

Sample questions

Keri plays cricket but never scores a century during the season.
  The number of runs scored by Keri in each game is shown in the following table:

Runs (r)	Frequency
$0 \leq r \leq 19$	4
$20 \leq r \leq 39$	12
$40 \leq r \leq 59$	8
$60 \leq r \leq 79$	4
$80 \leq r \leq 99$	2

For these data

1 find the mean number of runs Keri scores

2 write down the modal class.

Answers

1 Mean $= \dfrac{38 + 354 + 396 + 278 + 179}{30} = 41.5$.

Runs (r)	Frequency	Mid-interval value	Frequency × mid-interval value
$0 \leq r \leq 19$	4	9.5	38
$20 \leq r \leq 39$	12	29.5	354
$40 \leq r \leq 59$	8	49.5	396
$60 \leq r \leq 79$	4	69.5	278
$80 \leq r \leq 99$	2	89.5	179

2 Modal class $= 20 \leq r \leq 39$.

Now try this

3 A group of people take part in a sponsored walk. The times taken by each of them to complete the first mile of the walk are summarised in the following table. Calculate the mean time taken to complete the first mile.

Time (*t* min)	$0 \le t < 6$	$6 \le t < 12$	$12 \le t < 18$	$18 \le t < 24$	$24 \le t < 30$	$30 \le t < 36$
Frequency	0	4	13	16	6	1

4 A group of students guessed the height of a small tree. Their guesses are summarised in the following table. Calculate an estimate of the mean height guessed by these students.

Height (*x* m)	Number of students
$2.0 < x \le 2.2$	1
$2.2 < x \le 2.4$	14
$2.4 < x \le 2.6$	31
$2.6 < x \le 2.8$	29
$2.8 < x \le 3.0$	25
$3.0 < x \le 3.2$	17
$3.2 < x \le 3.4$	3

5 150 apples were picked and weighed. Their weights are shown in the following table. Calculate an estimate of the mean weight of an apple.

Weight (*x* g)	Number of apples (*f*)
$50 < x \le 60$	23
$60 < x \le 70$	34
$70 < x \le 80$	58
$80 < x \le 90$	20
$90 < x \le 100$	15
Total	**150**

● Sample exam question and answer

200 students were asked how far they travelled to school. The results are summarised in the following table. Calculate an estimate of the mean distance travelled.

Distance (*x* km)	Number of students
$0.0 < x \le 0.5$	40
$0.5 < x \le 1.0$	34
$1.0 < x \le 1.5$	55
$1.5 < x \le 2.0$	32
$2.0 < x \le 2.5$	23
$2.5 < x \le 3.0$	16

Distance (*x* km)	Number of students	Mid-interval value	Frequency × mid-interval value
$0.0 < x \le 0.5$	40	0.25	10.00
$0.5 < x \le 1.0$	34	0.75	25.50
$1.0 < x \le 1.5$	55	1.25	68.75
$1.5 < x \le 2.0$	32	1.75	65.00
$2.0 < x \le 2.5$	23	2.25	51.75
$2.5 < x \le 3.0$	16	2.75	44.00
Total	**200**		**265.00**

$$\text{Mean} = \frac{265}{200} = 1.325 \text{ km}$$

Exam-style questions

1 Alex used a spell checker to correct mistakes in his work. The following table shows the number of spelling mistakes on each page. Calculate the mean number of spelling mistakes per page.

Number of spelling mistakes on the page	0	1	2	3	4
Number of pages	10	6	1	1	2

2 The following table shows the number of screws in each of 100 boxes. Calculate the mean number of screws in a box.

Number of screws (x)	Number of boxes (f)
76	5
77	9
78	16
79	22
80	32
81	12
82	3
83	1
Total	**100**

3 The weights of 40 parcels in a small delivery office are summarised in the following table. Calculate an estimate of the mean weight of a parcel.

Weight (x kg)	Frequency (f)
$0 < x \le 2$	3
$2 < x \le 4$	9
$4 < x \le 6$	16
$6 < x \le 8$	8
$8 < x \le 10$	3
$10 < x \le 12$	1
$12 < x \le 14$	0

4 Jamilla has data on the length of time she spends on each telephone call she makes in 1 month. The data are summarised in the following table. None of her calls last longer than 10 minutes. Calculate an estimate of the mean length of time of a telephone call made by Jamilla.

Time (x min)	Frequency (f)
$0 < x \le 1$	4
$1 < x \le 2$	20
$2 < x \le 3$	12
$3 < x \le 4$	8
$4 < x \le 5$	17
$5 < x \le 6$	24
$6 < x \le 7$	36
$7 < x \le 8$	19
$8 < x \le 9$	8
$9 < x \le 10$	2

Collecting and displaying data

● Tally charts and frequency tables

Tally charts are a way of organising data. Tallies are bundled in fives. When tallies are added, they give the frequency of items in a data set.

Sample question

The scores in 25 different rounds of golf are shown below:

70	72	78	80	69	75	71	87
76	84	74	79	70	73	77	71
70	72	81	76	71	78	80	73	86

Draw a grouped frequency table to show these data using intervals of 66–70, 71–75, etc.

Answer

Score	Tally	Frequency
66–70	IIII	4
71–75	IIII IIII	9
76–80	IIII III	8
81–85	II	2
86–90	II	2

● Pictograms

Pictograms are ways of showing data using symbols. Each symbol represents a fixed number of data points.

Sample question

The following table shows information about the number of texts sent by a group of students. Each # represents two people.

Number of texts sent	
20	# # # #
21	# # # # # #
22	# # #
23	# # # #
24	# # # # #
25	# # # #
26	# # #
27	#

How many students sent 25 texts?

Answer

$4 \times 2 = 8$ students sent 25 texts.

● Pie charts

Pie charts are circles divided into sectors to show data. The sectors show the proportion of the data set not the frequency of items. The frequency of each sector can be calculated if you know the size of the whole data set.

Cambridge IGCSE Mathematics Study and Revision Guide Second Edition © Brian Seager *et al.*, 2016

Sample question

The following pie chart shows the make of mobile phones owned by 240 people.

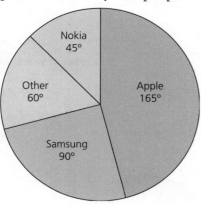

How many people own an Apple mobile?

Answer

Apple: $\frac{165}{360} \times 240 = 110$ people

Now try this

1 There are 40 weeks in a year when a school is open. A group of students was asked how many of those weeks they were never absent. The following data show their responses:

40   38   35   40   39   40   37   39   38   38   37   39   40
38   37   39   38   37   36   38   40   39   38   39   39   37
38   39   39   38

   a Draw a frequency table to show these data.
   b Draw a bar chart to show these data.

2 A class of students voted for their favourite tourist attraction. The votes are shown in the following table. Draw a pie chart to show these results.

Rockwall	2 votes
Canoe Chasm	6 votes
DeepDive	3 votes
The Wax House	7 votes
Adventure Dome	12 votes

3 A shop manager records the number of queries she receives each day about new internet connections. The data is summarised in the following pictogram:

Monday	⬭ ⌒
Tuesday	⬭ ⬭ ⬭
Wednesday	⬭ ⬭ ⌒
Thursday	
Friday	⬭ ⬭ ⬭
Saturday	⬭ ⬭ ⬭ ⬭ ⬭ ⌒

⬭ represents two queries.

   a How many queries were received on Saturday?
   b How many more queries were received on Friday than on Monday?
   c The total number of queries received in the week was 35. Complete the pictogram.

● Sample exam questions and answers

A small village recorded the number of sunny days each year it had over a period of 20 years. The results are listed below:

| 284 | 277 | 264 | 288 | 291 | 281 | 288 | 286 | 279 | 272 |
| 285 | 285 | 295 | 273 | 287 | 274 | 281 | 289 | 272 | 286 |

1 Work out the range of these values.

2 Complete the following frequency table:

3 Show this information in a pie chart.

Number of days		Number of years
261 to 270		
271 to 280		
281 to 290		
291 to 300		

1 $295 - 264 = 31$

2

Number of days		Number of years
261 to 270	I	1
271 to 280	IIIII I	6
281 to 290	IIIII IIIII I	11
291 to 300	II	2

3

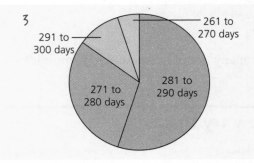

Exam-style questions

1 A group of students was asked how many different vowels there were in their first name. Their responses are shown below:

1 3 2 2 1 2 3 2 1 3 2 5 1 2

2 1 2 1 2 2 3 3 4 2 1 2 2 1

  a Complete the following frequency table of these results:

Number of different vowels	Frequency
1	8
2	
3	
4	
5	

  b Draw a bar chart to show this information.

2 The 720 students in a school were asked how they got to school in the morning. Their answers are summarised in the following table. Draw and label a pie chart to show this information.

Method of travel	Frequency
Walk	268
Cycle	140
Bus	196
Car	72
Other	44

3 The following pictogram shows the amount of their $2000 income a family allocate to various items.

Housing	⬭ ⬭ ⬭ ⬭ ⌒
Fuel	⬭ ⬭
Food	⬭ ⬭ ⬭ ⬭
Clothes	
Travel	⬭ ⬭ ⬭ ⌒
Other	⬭ ⬭ ⬭ ⬭ ⌒

⬭ represents $100.

  a How much do they spend on housing?

  b Complete the pictogram.

  c How much more do they spend on travel than on clothes?

Cambridge IGCSE Mathematics Study and Revision Guide Second Edition © Brian Seager *et al.*, 2016

● Surveys

Data can be collected using questionnaires. Some issues are:

- questions can give a choice of possible answers
- questions could be biased or leading
- questions should not cause embarrassment
- questions should be short, clear, easy to understand and relevant.

Sample questions

Here are two questions that Amy wanted to include in a survey she was going to do to find out if people liked the new shopping centre in her town.

Rewrite each one to show how you would word the questions in a questionnaire and state why you would change it.

1 How old are you?

2 This new shopping centre appears to be a success. Do you agree?

Answers

Possible answers would be:

1 The question may be thought to be personal — some people may not answer.
Change to: What is your age? Tick the appropriate box:
☐ 10–19 ☐ 20–29 ☐ 30–39
☐ 40–49 ☐ 50–59 ☐ Over 60

2 This is a leading question.
Change to: Do you think the shopping centre is a success? Tick the appropriate box:
☐ Yes ☐ No ☐ Do not know

Now try this

4 Bobby wants to choose a sample of 100 adults from the town where he lives. He considers these methods of choosing his sample:
Method 1 – choose people shopping in the town centre on Saturday
Method 2 – choose names at random from the list of electors
Method 3 – choose people living in streets near his home.
Which of these methods is most likely to provide an unbiased sample? Give reasons for your answer.

5 Colette wants to find out about people's TV viewing habits. Write down three questions Colette could ask.

● Sample exam questions and answers

Which of these questions do you think is/are biased?

1 It is important to eat breakfast. Do you eat breakfast?

2 Normal teenagers wear clothes with designer labels. Do you wear clothes with designer labels?

3 How many hours of television did you watch last weekend?

4 It is essential to eat five portions of fruit each day. How many portions did you eat yesterday?

1 Leading question, tells respondent they should be answering 'Yes'.
2 Leading question, tells the respondent they are not 'Normal' if they do not wear designer clothes.
3 Fair question. Probably better to offer categories of response to make the data manageable.
4 Leading question, tells respondent their answer should be five or more.

Exam-style questions

4 Here is a list of questions. Which of these questions do you think need(s) to change and why?

   a How heavy are you?

   b Where did you go on holiday last year?

   c How many newspapers have you read this week?

   d Do you enjoy sport?

   e How much money do you earn?

   f How many children do you have?

5 The Ayton Taxi Company attempted to estimate the number of people who used their taxis. They phoned 100 people in Ayton and asked if they had used one of their taxis in the last week. 19 people said 'yes', so the taxi company claimed that '19% of the people of Ayton use our taxis'. Give three reasons why this was not a good method of estimating customer usage.

● Scatter diagrams and lines of best fit

Scatter diagrams are a means of establishing if a relationship (correlation) exists between two variables. Where a relationship does exist, it is possible to draw a line of best fit that can be used to estimate one value given the other.

Correlation can be strong (all points close to a line) or weak (points more spread out but a visible pattern).

Correlation can also be positive (as one value increases, so does the other) or negative (as one value increases the other decreases). Where all the data are widely spread and there is no pattern, there is no correlation.

Sample questions

The following table shows the height (cm) and the mass (kg) of a group of students at a high school.

Height	170	155	163	181	157	169	172
Mass	56	47	61	72	57	64	63

Height	187	168	177	159	164	183	188
Mass	70	65	62	53	58	68	75

Height	165	172	178	184	167	181	163
Mass	62	66	62	73	61	64	64

Height	190	172	184	159			
Mass	72	67	71	45			

1 Draw a scatter diagram to show this information.

2 Describe the correlation.

3 Draw a line of best fit for the data.

4 Another student is 175 cm tall. Use your diagram to estimate the mass of this student.

Answers

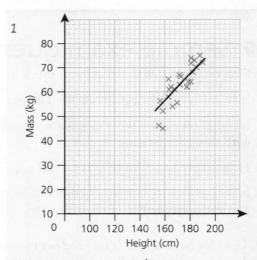

2 Weak positive correlation.
3 See diagram.
4 65 kg.

Cambridge IGCSE Mathematics Study and Revision Guide Second Edition © Brian Seager *et al.*, 2016

Now try this

6 The following table shows the height and shoe size of seven students:

Height (cm)	165	152	145	161	148	158	156
Shoe size	6	5	$3\frac{1}{2}$	$5\frac{1}{2}$	4	$5\frac{1}{2}$	5

   a Draw a scatter diagram. Describe the correlation.
   b Draw a line of best fit.
   c Estimate the shoe size for a student of height 151 cm.

7 Market research suggests that the following sales (in thousands) would result from pricing an item at the given price:

Price ($)	35	40	45	50	55	60	65
Sales (thousands)	25	24	23	20	16	14	11

   a Draw a scatter diagram. Describe the correlation.
   b Draw a line of best fit.
   c Estimate the sales if the price was $52.

● Sample exam questions and answers

The following table shows the time spent and marks gained for ten students' coursework:

Time (h)	5	8	3	6	6	7	4	10	8	7
Mark	12	15	9	14	11	13	10	19	16	14

1 Plot a scatter diagram for this information.

2 What conclusions can be drawn?

3 Draw a line of best fit.

4 Use your line of best fit to estimate the mark of a student who spent nine hours on the coursework.

1 and 3

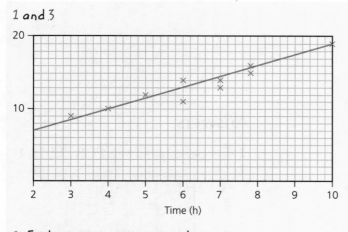

2 Fairly strong position correlation.
4 17 or 18.

Exam-style questions

6 An orchard contains nine young apple trees. The following table shows the height of each tree and the number of apples on each:

Height (m)	1.5	1.9	1.6	2.2	2.1	1.3	2.6	2.1	1.4
Number of apples	12	15	20	17	20	8	26	22	10

a Draw a scatter graph to illustrate this information. Use a scale of 2 cm to 1 m on the horizontal axis and 2 cm to 10 apples on the vertical axis.

b Comment briefly on the relationship between the height of the trees and the number of apples on the trees.

c Add a line of best fit to your scatter graph.

d Explain why it is not reasonable to use this line to estimate the number of apples on a tree of similar type but of height 4 m.

7 Bob thinks there is a relationship between arm length and the length of the first finger of the writing hand. He measured these lengths for 14 students in his class. The measurements, in centimetres, are given in the following table:

Arm	56	65.5	73.5	64	64	81	57	60	69	53	65	72.5	57.5	74.5
Finger	6	6.5	7.5	7	6	8	6.5	6.5	7	6	7	7.5	7	7.5

a Draw a scatter graph to show these data.

b What does the graph show about any connection between the two lengths?

c Bob knows the arm length for a student is 68 cm. What would you expect the finger length to be?

● Histograms

Histograms are particularly useful when you are dealing with unequal intervals.

A histogram is similar to a frequency diagram, except that, on a histogram, the area of the bar is equal to the frequency.

● Width of interval × height of bar = frequency

● Height = $\dfrac{\text{frequency}}{\text{width of interval}}$

The height is called the frequency density.

Sample questions

The following table shows the distribution of times (t seconds) that a company took to answer 100 telephone calls.

Time (t s)	Frequency (f)	Frequency density = $f \div$ width
$0 < t \leq 2$	28	
$2 < t \leq 5$	48	
$5 < t \leq 10$	12	
$10 < t \leq 15$	7	
$15 < t \leq 20$	5	

1 Complete the column for frequency density.

2 Draw a histogram to represent the data.

Answers

1 Frequency densities: 14, 16, 2.4, 1.4, 1.

2

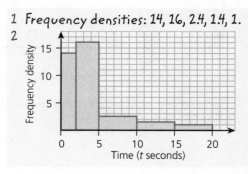

Cambridge IGCSE Mathematics Study and Revision Guide Second Edition © Brian Seager *et al.*, 2016

Now try this

8 The distribution of the masses (*m*, kg) of 200 parcels received at a parcel sorting office is shown in the following table:

Mass (*m* kg)	Frequency (*f*)	Frequency density = *f* ÷ width
$0 < m \leq 0.5$	21	
$0.5 < m \leq 1$	33	
$1 < m \leq 2$	55	
$2 < m \leq 5$	45	
$5 < m \leq 10$	30	
$10 < m \leq 20$	16	

a Copy the table and complete the column for frequency density.
b Draw a histogram to represent the data.

9 The following table shows the distribution of heights (*h*, cm) of 120 plants:

Height (*h* cm)	Frequency (*f*)	Frequency density = *f* ÷ width
$0 < h \leq 5$	8	
$5 < h \leq 10$	20	
$10 < h \leq 20$	50	
$20 < h \leq 30$	25	
$30 < h \leq 50$	17	

a Copy the table and complete the column for frequency density.
b Draw a histogram to represent the data.
c Calculate the mean height of the plants.

● Sample exam question and answer

A test was carried out to establish the ability of a mouse to find food. The test was carried out on 120 mice. The distribution of times taken to reach the food is given in the following table:

Draw a histogram to represent this information.

Time (*t* s)	Frequency
$0 < t \leq 10$	18
$10 < t \leq 15$	46
$15 < t \leq 20$	35
$20 < t \leq 30$	13
$30 < t \leq 50$	8

Time (*t* s)	Frequency	Frequency density
$0 < t \leq 10$	18	1.8
$10 < t \leq 15$	46	9.2
$15 < t \leq 20$	35	7
$20 < t \leq 30$	13	1.3
$30 < t \leq 50$	8	0.4

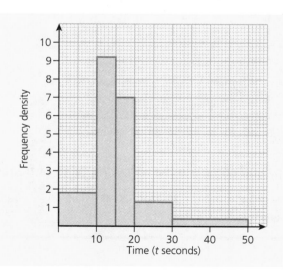

Exam-style questions

8 A doctor's patients are divided by ages as shown in the following table:

Age (x) in years	$0 \leq x < 5$	$5 \leq x < 15$	$15 \leq x < 25$	$25 \leq x < 45$	$45 \leq x < 75$
Number of calls	14	41	59	70	16

Draw a histogram to represent this information. Use a scale of 2 cm to 10 years on the x-axis and an area of 1 cm² to represent five patients.

9 A sample was taken of the telephone calls to a school switchboard. The lengths of the telephone calls are recorded, in minutes, in the following table:

Time (t min)	$0 < t \leq 1$	$1 < t \leq 3$	$3 < t \leq 5$	$5 < t \leq 10$	$10 \leq t \leq 20$
Number of calls	12	32	19	20	15

Copy and complete the histogram to show this information.

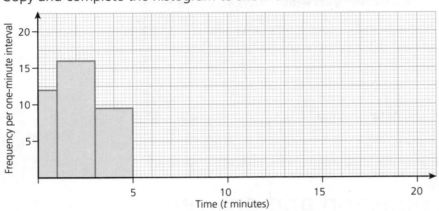

10 All of the pupils in a small secondary school were asked how long, on average, they spent each night completing their homework. The results are shown in the following histogram:

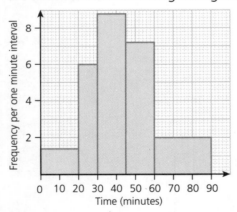

Copy and complete the following table of frequencies:

Time (min)	0–20	20–30	30–45	45–60	60–90
Frequency		60			

(39) Cumulative frequency

Key objectives

- To construct and use cumulative frequency diagrams.
- To estimate and interpret the median, percentiles, quartiles and inter-quartile range.

● Cumulative frequency graphs

Cumulative frequency is the running total of the frequencies in a distribution. The final cumulative frequency is the total of the frequencies and is often given in the question. Cumulative frequency is plotted at the **upper bound of each interval**. The points can be joined by a curve or by straight lines. The graph in the Sample question below shows a typical shape for a cumulative frequency curve. To find the median of a frequency distribution, draw a line across the graph at half the total frequency to meet the curve and then down to read off the value on the x-axis.

● Quartiles and the inter-quartile range

The cumulative frequency axis can also be represented as **percentiles.** The **lower quartile** would be at the 25th percentile (or one-quarter of the cumulative frequency) and the **upper quartile** would be at the 75th percentile (or three-quarters of the cumulative frequency). To find the quartiles of a frequency distribution, draw lines across the graph at one-quarter (25%) and three-quarters (75%) of the total frequency to meet the curve and then down to read off the values on the x-axis.

The **inter-quartile range** (upper quartile–lower quartile) is a measure of spread.

Sample questions

The following table shows the distribution of the masses (m grams) of 80 tomatoes:

Mass (m g)	Frequency	Cumulative frequency
$70 < m \leq 75$	4	
$75 < m \leq 80$	10	
$80 < m \leq 85$	22	
$85 < m \leq 90$	27	
$90 < m \leq 95$	14	
$95 < m \leq 100$	3	

1 Complete the right-hand column to show the cumulative frequencies.

2 Draw a cumulative frequency diagram.

3 Estimate the number of tomatoes weighing more than 93 g.

4 Find the median and inter-quartile range.

Answers

1

Mass (*m* g)	Frequency	Cumulative frequency
70<*m*≤75	4	4
75<*m*≤80	10	14
80<*m*≤85	22	36
85<*m*≤90	27	63
90<*m*≤95	14	77
95<*m*≤100	3	80

2

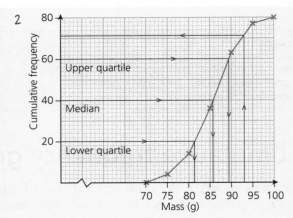

3 $80 - 71 = 9$

4 Median $= 86\,g$
Inter-quartile range $= 89.5 - 81.5 = 8\,g$

Now try this

1 The following table shows the results of a survey on the times (*t* minutes) taken by 100 students to get to school:

Time (*t* min)	Frequency
0<*t*≤10	8
10<*t*≤20	15
20<*t*≤30	33
30<*t*≤40	22
40<*t*≤50	17
50<*t*≤60	5

a Draw up a table of cumulative frequencies.
b Draw a cumulative frequency graph.
c Use your graph to find
   (i) the number of students who took over 45 min to get to school
   (ii) the median time
   (iii) the inter-quartile range.

2 The following cumulative frequency graph shows the marks obtained by 120 students in their last mathematics examination:

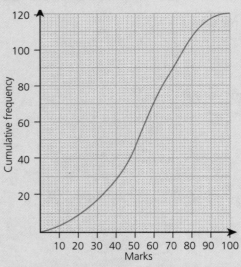

a The pass mark was 45. How many students passed?
b Find the median and inter-quartile range.

3 50 students completed a puzzle. The time taken by each student was recorded and the following table shows a summary of the data:

Time (*t* min)	0<*t*≤5	5<*t*≤10	10<*t*≤15	15<*t*≤20	20<*t*≤25	25<*t*≤30	30<*t*≤35
Number of students	4	12	16	9	6	2	1

a Draw a cumulative frequency curve for these data.
b Use your diagram to find the median time taken.

Cambridge IGCSE Mathematics Study and Revision Guide Second Edition © Brian Seager *et al.*, 2016

4 The following table shows the weights, in kilograms, of 80 sheep on a farm:

Weight (w kg)	$40 < w \le 50$	$50 < w \le 60$	$60 < w \le 70$	$70 < w \le 80$	$80 < w \le 90$	$90 < w \le 100$
Frequency	5	7	16	24	19	9

a Complete the following cumulative frequency table:

Weight (w kg)	$w \le 50$	$w \le 60$	$w \le 70$	$w \le 80$	$w \le 90$	$w \le 100$
Cumulative frequency	5					

b Draw a cumulative frequency diagram for this data.
c Use your diagram to find the inter-quartile range of the weights of the sheep.
d All the sheep with a weight greater than 72 kg are sent to market. Estimate the number of sheep sent to market.

● Sample exam questions and answers

A test was carried out to establish the ability of a mouse to find food. The test was carried out on 120 mice. The distribution of times taken to reach the food is given in the following table:

Time (t s)	Frequency
$0 < t \le 5$	18
$5 < t \le 10$	46
$10 < t \le 15$	35
$15 < t \le 20$	13
$20 < t \le 25$	8

1 Draw a cumulative frequency graph to represent this information.

2 Find the median time it takes a mouse to find food.

3 Find the inter-quartile range of the time it takes a mouse to find food.

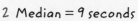

1

Time (t s)	Frequency
$t \le 5$	18
$t \le 10$	64
$t \le 15$	99
$t \le 20$	112
$t \le 25$	120

2 Median = 9 seconds
3 Inter-quartile
   range = 13.5 − 7.5 = 6

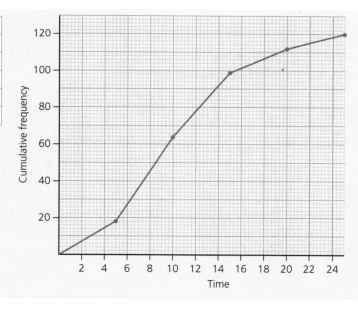

Exam-style questions

1 The weight in kilograms of each of 80 children under 10 years in a village is summarised in the following table:

Weight (w kg)	Frequency
$0 < w \leq 5$	9
$5 < w \leq 10$	13
$10 < w \leq 15$	21
$15 < w \leq 20$	17
$20 < w \leq 25$	10
$25 < w \leq 30$	8
$30 < w \leq 35$	2

a (i) Copy and complete the following cumulative frequency table:

Weight (w kg)	Cumulative frequency
$w \leq 5$	9
$w \leq 10$	
$w \leq 15$	
$w \leq 20$	
$w \leq 25$	
$w \leq 30$	
$w \leq 35$	

(ii) Draw the cumulative frequency graph for the weight of the children. Use a scale of 2 cm to 5 kg on the horizontal axis and 1 cm to 10 children on the vertical axis.

b Use your graph to find

(i) the median weight

(ii) the inter-quartile range.

c Use your graph to estimate how many children weigh at least 8 kg.

2 Yasmin recorded the playing times of each track in her CD collection. The following table shows the grouped distribution times:

Time (t, min)	$1 < t \leq 2$	$2 < t \leq 3$	$3 < t \leq 4$	$4 < t \leq 5$	$5 < t \leq 6$	$6 < t \leq 7$
Number of tracks (frequency)	5	25	45	82	33	10

a (i) Copy and complete the following cumulative frequency table:

Time (t, min)	$t \leq 1$	$t \leq 2$	$t \leq 3$	$t \leq 4$	$t \leq 5$	$t \leq 6$	$t \leq 7$
Number of tracks	0	5					200

(ii) Draw the cumulative frequency curve showing the playing times of the CD tracks. Use a scale of 2 cm to 1 min on the horizontal axis and 2 cm to 50 tracks on the vertical axis.

b Use the cumulative frequency curve to estimate

(i) the median playing time of a CD track

(ii) the probability that a randomly chosen track plays for longer than 3.7 min.

Cambridge IGCSE Mathematics Study and Revision Guide Second Edition © Brian Seager *et al.*, 2016

3 A ferry company conducted a survey of 80 vehicles waiting for the noon ferry and asked the drivers how long, in minutes, they waited. The following table shows a summary of the data:

Time (*t* min)	$0<t\leq20$	$20<t\leq40$	$40<t\leq60$	$60<t\leq80$	$80<t\leq100$
Number of drivers (frequency)	4	19	30	18	9

a Complete the following cumulative frequency table:

Time (*t* min)	$t\leq20$	$t\leq40$	$t\leq60$	$t\leq80$	$t\leq100$
Cumulative frequency	4				

b Draw a cumulative frequency diagram for these data.

c Use your diagram to find

  (i) the median waiting time

  (ii) the inter-quartile range of the waiting times.

Answers

● Chapter 1

Now try this

1 a 4, 8, 12, 16, 20, 24

  b 9, 18, 27, 36, 45, 54

  c 15, 30, 45, 60, 75, 90

2 a 1, 2, 3, 4, 6, 8, 12, 24

  b 1, 2, 4, 5, 8, 10, 20, 40

  c 1, 2, 4, 8, 16, 32

3 a $2 \times 2 \times 3 \times 3$

  b $2 \times 2 \times 5 \times 7$

  c $2 \times 2 \times 3 \times 7$

4 a HCF = 3, LCM = 120

  b HCF = 4, LCM = 120

  c HCF = 18, LCM = 540

5 a $30   **b** Six floors

6 a 1600   **b** 9   **c** 216   **d** 7

7 $\frac{\pi}{6}$ and $\sqrt[3]{17}$

Exam-style questions

1 a 14   **b** 12 or 18

  c 11, 13, 17 or 19

2 $264 = 2 \times 2 \times 2 \times 3 \times 11$

3 HCF = 4, LCM = 48

4 HCF = 2, LCM = 60

5 1

6 13

● Chapter 2

Now try this

1 a 30.97   **b** 4.14   **c** 24.60

2 a 5350   **b** 61.4   **c** 3050

3 a $800 \div 20 = 40$ **b** $\frac{6000}{60 \times 20} = 5$

4 a 72.5 to 73.5 g

  b 4.25 to 4.35 litres

  c 7.045 to 7.055 m

5 25.65 to 25.75 s

6 Lower = 3.67, upper = 4.62

7 3.11 to 3.20 cm

Exam-style questions

1 a 4 410 000 mm^3 **b** 0.00441 m^3

2 9 kg

3 5.10

4 Lower = 43.6 people/km^2, upper = 46.1 people/km^2

5 3246 pools

● Chapter 3

Now try this

1 a $x = -4, -3, -2, -1$

  b $x = 2, 3, 4, 5$

  c $x = 2, 3, 4$

  d $x = -4, -3, -2, -1$

2 $\frac{1}{3}, \frac{1}{2}, \frac{2}{3}, \frac{5}{6}$

3 $\frac{3}{8}, \frac{1}{2}, \frac{5}{8}, \frac{3}{4}$

4 $\frac{1}{4}, \frac{5}{12}, \frac{7}{12}, \frac{2}{3}$

5 $\frac{7}{10}, 72\%, 0.73, \frac{3}{4}$

6 49%, 0.5, 53%, $\frac{3}{5}$, 0.65

7 33%, $\frac{7}{20}$, 0.38, $\frac{2}{5}$, 0.42

8 a 416   **b** 2016   **c** 288

9 a 544   **b** 720   **c** 2288

10 a 62   **b** 70   **c** 17

11 a 43.35   **b** 8.04   **c** 25.26   **d** 4.00

Exam-style questions

1 $\frac{7}{16}, \frac{1}{2}, \frac{5}{8}, \frac{3}{4}$

2 55%, 0.59, $\frac{3}{5}$, 0.62, 66%

3 a 3.49   **b** 0.088   **c** 14.64

4 a 831.08

  b 5.36

Cambridge IGCSE Mathematics Study and Revision Guide Second Edition © Brian Seager *et al.*, 2016

● Chapter 4

Now try this

1 a 34%   b 27%   c 3%   d 42.8%

2 a (i) 0.7   (ii) 70%
  b (i) 0.6   (ii) 60%
  c (i) 0.45   (ii) 45%

3 a (i) 0.4   (ii) $\frac{2}{5}$
  b (i) 0.24   (ii) $\frac{6}{25}$
  c (i) 0.05   (ii) $\frac{1}{20}$

4 $5\frac{13}{24}$ 5 $5\frac{7}{15}$ 6 $1\frac{11}{20}$
7 $1\frac{1}{12}$ 8 $2\frac{17}{30}$ 9 $\frac{23}{24}$
10 $10\frac{1}{2}$ 11 4   12 $11\frac{2}{5}$
13 $3\frac{1}{3}$ 14 $6\frac{2}{3}$ 15 $1\frac{4}{5}$
16 $\frac{7}{9}$ 17 $\frac{61}{99}$ 18 $\frac{5}{12}$

Exam-style questions

1 a $3\frac{1}{8}$ b $1\frac{1}{6}$
2 a $3\frac{5}{9}$ b $3\frac{17}{20}$
3 a $22   b $90
4 a $\frac{7}{8}$ b $1\frac{1}{4}$
5 $h = \frac{1}{3}$
6 18460

● Chapter 5

Now try this

1 a $7.20   b 17.02 m
2 35%   3 12.5%
4 60%   5 12.5%
6 a $89.88   b $6457.50
7 a $10.80   b $5723
8 $1714.75   9 $191896.50
10 60   11 $8200
12 $1200   13 $8120

Exam-style questions

1 $8.17   2 84.7%
3 a $68   b $38
4 a $763.75   b 14.9%

● Chapter 6

Now try this

1 a 0.5 hrs   b 4 hrs   c 5 hrs
2 440 m

3 a (i) 1:3   (ii) 8:3
    (iii) 1:5   (iv) 2:5
  b (i) 1:3   (ii) 1:0.375
    (iii) 1:5   (iv) 1:2.5

4 $8, $10   5 21 litres
6 $29.50, $59, $88.50   7 12 minutes
8 4 weeks   9 55 minutes
10 9 days
11 a 80   b 54   c 57.6
12 a 35   b 37.5   c 31.25

Exam-style questions

1 a 160 g   b 150 biscuits
2 a 100 ml   b 300 litres
3 a 17.5 cm   b 1:40 000
4 a 70%   b 140 men and
                160 women
5 16 kilometres per hour

● Chapter 7

Now try this

1 81   2 1
3 $\frac{1}{9}$ 4 1
5 216   6 $\frac{1}{8}$
7 5^{10} 8 6^3
9 3^8 10 2^{-6}
11 2^9 12 5^5
13 a 2   b −3
  c 1   d −1
14 372 000   15 0.00048
16 5.83×10^{-5} 17 7.56×10^7
18 a 4.6×10^4 b 4.84×10^{-2}
  c 3.6×10^6 d 5.965×10^3

19 a 8 000 000   **b** 9400

    c 77   **d** 0.000303

20 3×10^{-13} **21** 2.4×10^5

22 5×10^{10} **23** 4×10^{-1} or 0.4

24 1.2×10^{14} **25** 2.5×10^{10}

26 6   **27** 0.1

28 20.1   **29** 0.0625

30 a 12   **b** $\frac{1}{9}$

    c 0.5   **d** 1

31 $4^{-2}, 4^0, 4^{\frac{1}{3}}, 4^{\frac{1}{2}}, 4^2$

Exam-style questions

1 a 2^{11} **b** 2.05×10^3

2 2.4×10^{10}

3 a 1.8×10^8 **b** 1.2×10^{-4}

4 8.75×10^{-3}

5 a $\frac{1}{81}$ **b** 1   **c** 3

6 a (i) 1   **(ii)** 3

    b 2.4×10^4

7 $\frac{1}{2}$

● Chapter 8

Now try this

1 $16, $24, $32

2 $467.70

3 a 2875 rands   **b** $69.57 to the nearest cent

4 $4

5 a $340   **b** $1360

6 a $57.60   **b** $381.13 to nearest cent

7 4.5%

8 $3130.39 to nearest cent

9 $750 (after rounding)

10 a 336.0   **b** 0.174   **c** 29%

11 4660 to 3 s.f.

12 $4331.93

13 11975

Exam-style questions

1 Special

2 a $24.50   **b** $7.32

3 a $4900.17   **b** 8 years

4 $750

5 30.2 litres

● Chapter 9

Now try this

1 a 0150   **b** 1440

    c 1149   **d** 1830

2 a 3:45 a.m.   **b** 2:56 p.m.

    c 11:40 p.m.   **d** 11:59 a.m.

3 7:55 a.m.

4 12 km/h

5 12:49 p.m.

Exam-style questions

1 6 hours 47 minutes

2 64.3 km h^{-1}

● Chapter 10

Now try this

1 $\{10, 12, 14, 16, 18, 20\}$

2 When $x = 4$, $y = 2 \times 4 - 7 = 1$, so $(4, 1) \in \{(x, y): y = 2x - 7\}$

3

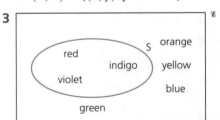

4 a $\{1, 4, 5, 7, 8, 9, 10\}$

    b (i) $\in$ **(ii)** $\notin$ **(iii)** $\in$

5 a $\{3\}$ **b** $\{ \}$, or $\varnothing$

    c $\{1, 2, 3, 5, 6\}$

    d $\{4, 6, 7\}$

6

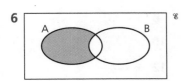

7 { }, {a}, {b}, {c}, {a, b}, {a, c}, {b, c}, {a, b, c}

Exam-style questions

1

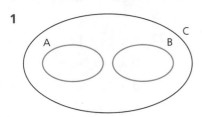

2 a {6, 12}   **b** {1, 5, 7, 11}

3 Total = 5 + 6 + 9 + 10 = 30

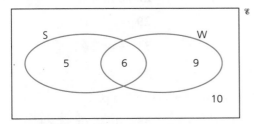

4 a

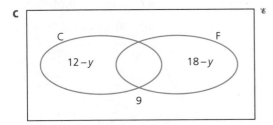

b S′ ∪ T

c

C   F

12 − y   18 − y

9

5 students study chemistry but not French.

● Chapter 11

Now try this

1 3

2 a 7.125   **b** $-1\frac{7}{8}$ or −1.875

3 25

---

4 $4x + 8 = 40$ Width = 8 cm, length = 12 cm.

5 $2x + 10 = 58$ Women = 24, men = 34.

6 $x^2 + 8 = 89$ Number = ±9.

7 $c = y - mx$ **8** $m = \dfrac{y - c}{x}$

9 $b = \dfrac{x + cd}{a}$ **10** $c = \dfrac{ab - x}{d}$

11 $w = \dfrac{P - 2i}{2}$ or $\dfrac{P}{2} - i$ **12** $h = \dfrac{3V}{iw}$

13 $r = \sqrt{\dfrac{A}{4\pi}}$ **14** $x = \sqrt{y - a}$

15 $x = \dfrac{(y - ay)}{a}$ **16** $r = \sqrt[3]{\dfrac{3V}{4\pi}}$

17 $y = \dfrac{ax}{1 - b}$ **18** $c = \dfrac{ab + bd}{a - d}$

19 $8a^2 + a$ **20** $2x^2 + 21x$

21 $6x - 14y$ **22** $2x^2 - x$

23 $x^2 + 9x + 20$ **24** $x^2 + 4x - 21$

25 $x^2 - x - 2$ **26** $x^2 - 12x + 27$

27 $2x^2 + 4x - 6$ **28** $6x^2 - 7x + 2$

29 $5x^2 - 11x - 12$ **30** $a^2 - ab - 6b^2$

31 $y^2 - 4$ **32** $16x^2 + 40x + 25$

33 $5(x - 5)$ **34** $a(3b - 2c)$

35 $2ab(a - 3b)$ **36** $2a(2bc - 4b + 1)$

37 $(5 + y)(2x + 3)$ **38** $(2 + 3y)(x + 3)$

39 $(4 + 3y)(2x + 3)$ **40** $(4 + 3a)(b + 5)$

41 $(7a + 3)(2 + b)$ **42** $(3a + 2)(2 + 3b)$

43 $(3 + 2y)(x - 5)$ **44** $(3 + 2y)(x - 4)$

45 $(2 - 3y)(3x + 4)$ **46** $(2 - a)(3b + 4)$

47 $(x - y)(x + y)$ **48** $(r - 4s)(r + 4s)$

49 $(p - q)(p + q)$ **50** $2(a - 2b)(a + 2b)$

51 $(v - w)(v + w)$ **52** $4(2m - 3n)(2m + 3n)$

53 $(x - 2y)(x + 2y)$ **54** $5(x - 2y)(x + 2y)$

55 $(3a - b)(3a + b)$ **56** $(x^3 - y)(x^3 + y)$

57 $(x + 5)(x + 1)$ **58** $(x - 5)(x + 3)$

59 $(3x - 2)(x - 1)$ **60** $(x - 4)(x - 2)$

61 $(x + 3)(x - 3)$ **62** $(x + 9)(x - 9)$

63 $3(x + 3)(x - 3)$ **64** $\dfrac{x - 2}{x + 1}$

65 $\dfrac{2(x + 2)}{2x + 1}$ **66** $\dfrac{3x - 1}{x(x - 1)}$

67 $\dfrac{3(2x - 1)}{(x + 1)(x - 2)}$

Cambridge IGCSE Mathematics Study and Revision Guide Second Edition © Brian Seager *et al.*, 2016

Exam-style questions

1 $n = \dfrac{P - 120}{4}$

2 144

3 a 1432.2   b $y = \sqrt{\dfrac{x+3}{4}}$

4 a $d = \dfrac{e-3}{5}$ b $d = \dfrac{4e+7}{3-5e}$

5 a −32.3   b $v = \dfrac{fu}{u-f}$

6 $3x + 2(x+45) = 415$. Small radiator $65

7 $3pq(4p - 5q)$

8 a $2x + 14$ b $9 - 6x + x^2$

9 a $3x^2 + 10x - 8$ b $2y^2 - 7y + 3$

10 a $(5a-3)(3-2b)$ b $(a-2)(3-5b)$

11 a $2(5a-3b)(5a+3b)$
    b $(x-y)(x+y)(x^2+y^2)$

12 a $x(x+1)$ b $6(x-2)(x+2)$
    c $(x-4)(x+2)$ d $(2x-3)(x-15)$

13 $\dfrac{x}{x-2}$

14 a $\dfrac{x+3}{x+2}$ b $\dfrac{-3x+7}{(3x+1)(x+1)}$

Chapter 12

Now try this

1 8   2 9

3 5   4 8

5 $8a^2b^3c^3$ 6 $64a^6$

7 $\dfrac{7b^2}{a}$ 8 $\dfrac{6b}{a}$

Exam-style questions

1 a $8pq^3$ b $6a$

2 a $12a^3b^4$ b $7a^2b^{-1}$

3 a $\frac{1}{2}$ b $\dfrac{p^3}{q}$

Chapter 13

Now try this

1 $x = 5$ 2 $p = 2$

3 $m = 2\frac{1}{4}$ 4 $y = 8$

5 $m = 9$ 6 $p = 2$

7 $x = 8$ 8 $x = -3$

9 $x = 3$ 10 $x = 2\frac{2}{5}$

11 $x = \frac{1}{5}$ 12 $x = -4$

13 $x = -18$ 14 $m = 23$

15 $p = 42$ 16 $x = 11$

17 $x = 16$ 18 $x = 8$

19 $x = 1$ 20 $x = 8$

21 $x = 5$ 22 $x = \dfrac{24}{5} = 4\frac{4}{5}$

23 a $x = 3, y = 2$ b $x = 6, y = -5$
    c $x = 2, y = -1$ d $x = 1, y = -2$
    e $x = 2\frac{1}{2}, y = -3\frac{1}{2}$ f $x = 4, y = -3$

24 3 or 6   25 −4 or 1

26 −4 or −1   27 3 or 4

28 2 or −5   29 −0.5 or 7

30 2 or $-\frac{5}{3}$ 31 $\frac{1}{3}$ or $2\frac{1}{2}$

32 0 or −4   33 5 or −5

Answers 34 to 49 are correct to 2 d.p.

34 5.16 or −1.16   35 −0.26 or −7.74

36 9.24 or 0.76   37 2.62 or 0.38

38 1.87 or 0.13   39 3.12 or −1.12

40 0.16 or −4.16   41 −0.14 or −5.86

42 4.30 or 0.70   43 −0.13 or −3.87

44 5.24 or 0.76   45 1.85 or −0.18

46 1.39 or 0.36   47 5.18 or −0.68

48 −0.17 or −3.83   49 −0.21 or −3.79

50

There are no real solutions, as you cannot find the square root of a negative number.

The graph does not cross the line $y = 0$.

51 a $x < 2$ b $x < -\frac{3}{5}$ c $x \geq \frac{1}{6}$

52 $x > 1$, so smallest value is 2

53 $x \geq 5$ and $x \leq -5$

54 $-6 < x < -2$

Cambridge IGCSE Mathematics Study and Revision Guide Second Edition © Brian Seager *et al.*, 2016

55 a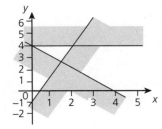

b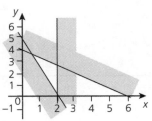

Exam-style questions

1 $x = \frac{1}{2}$

2 $p = 6$

3 $m = 12$

4 $y = 2$

5 $x = 3.5$

6 $x = 1$

7 a $4x + 4 = 36$ **b** $x = 8\,cm$

   c Area $= 80\,cm^2$

8 $x = 3, y = -\frac{1}{2}$

9 $x = -1, y = 3$

10 a By substituting $x = 3$ and $y = 2$

   b $9p + 11q = 5$ **c** $p = 3, q = -2$

11 a $x = 2$ or 4 **b** $x = 1\frac{1}{2}$ or -3

12 a $5(x+2)(x-2)$

   b (i) $(x-8)(x-1)$ **(ii)** $x = 8$ or 1

13 a $6x^2 + 9x - 42$

   b (i) $x(x+6)$ **(ii)** $x = 0$ or -6

14 $x = 17.73$ or 1.27 to 2 d.p.

15 a $3(x-2)^2 - 10$

   b $x = 3.83$ or 0.17 to 2 d.p.

16 a $y(15 - y) = 55$

           $15y - y^2 = 55$

       $y^2 - 15y + 55 = 0$

   b Length $= 8.62$ and width $= 6.38\,cm$

17 a Length (in m) $= 22 - 2x$

     Area (in m²) $= x(22 - 2x) = 60$

     $22x - 2x^2 = 60$, $11x - x^2 = 30$, $x^2 - 11x + 30 = 0$

   b $x = 6$ or 5

   c The pen is 6 m by 10 m or 5 m by 12 m

18 $x > 2.5$

19 $x < 5.5$

20 $x < 6$ and $x > -6$ or $-6 < x < 6$

21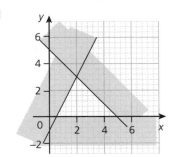

22 $x + y \le 3$

       $y \ge x$

       $x \ge -2$

● Chapter 14

Now try this

1 $x \ge 5, y \ge 5, x + y < 15$

2 $y > x, x + y \le 600, 5x + 8y \ge 3200$

3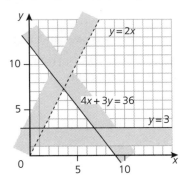

4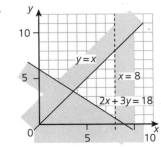

5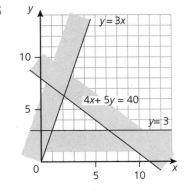

6 a The points (4, 6), (3, 5), (4, 5), (5, 5), (3, 4),
(4, 4), (5, 4), (6, 4), (2, 3), (3, 3), (4, 3), (5, 3)
and (6, 3) marked

b (i) (4, 6)   **(ii)** 16

7 a (6, 2)   **b** 18

8 (3, 3), (4, 4)

Exam-style questions

1 a $x + 2y \leq 28$, $3x + y \leq 24$

b
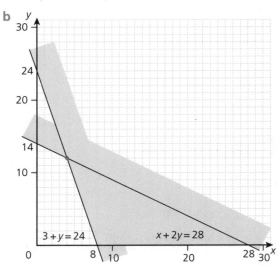

c (i) \$200   **(ii)** 4 type A and 12 type B

2 a (i) \$1000 is available, so $10x + 25y \leq 1000$.
Dividing by 5 gives $2x + 5y \leq 200$

(ii) $y \geq x$ and $x + y \geq 50$

b
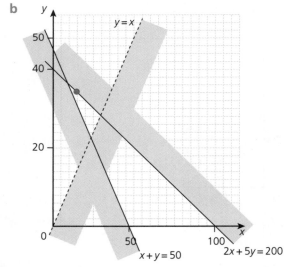

c 33

3 a $x > y$, $x + 3y \geq 30$

b
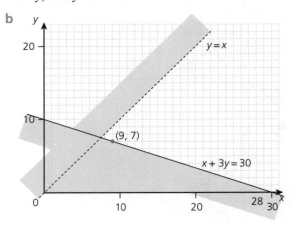

c (i) 23 hours

(ii) Nine individual photos and seven sets

● Chapter 15

Now try this

1 6, 9, 12, 15   **2** Add 5

3 89, $6n - 1$ **4** $44 - 4n$

5 $2n + 1$ **6** $10n$

7 $5n - 9$ **8** $35 - 3n$

9 a $2 \times 4^{n-1}$ **b** 5^{n-1} **c** $(n+2)(n+4)$

Exam-style questions

1 248, $5n - 2$

2 $5n - 4$

3 a $7n - 5$

b If $7n - 5 = 300$, $7n = 305$, $n = 305 \div 7$, $n = 43.57$.
n is not a whole number, so 300 is not in
the sequence

4 a n^2 **b** $3n^2 + 1$

5 a $2n + 1$ **b** $2n^2 + n + 1$

6 a n^3 **b** $2n^3$ **c** $2n^3 - 3n$

● Chapter 16

Now try this

1 192   **2** 90

3 43.75 litres   **4** 240.1 m

5 1

6 a 6   **b** 16

7 0.5 m   **8** 0.2 lumens

Exam-style questions

1 300   **2** 2000

3 72 m

● Chapter 17

Now try this

1 a Accelerated at 0.8 m/s² for 25 seconds to 20 m/s. Constant speed of 20 m/s for 15 seconds. Decelerated at 2 m/s² for 10 seconds to stop

     b Single tap filling at 4 litres/minute for 5 minutes followed by both taps filling at 8 litres/minutes for $2\frac{1}{2}$ minutes. Constant at 40 litres for $12\frac{1}{2}$ minutes. Emptied at 8 litres/minute in 5 minutes

2

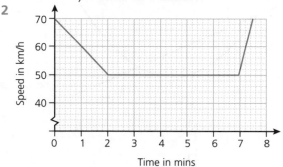

3 550 m

4

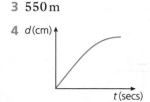

Exam-style questions

1 a

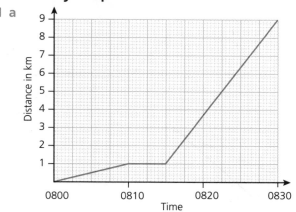

b $3\frac{2}{3}$ km or 3.7 km

2 For example: A car accelerates and then travels at a steady speed and then stops suddenly (crashes, etc.)

3 a   **b**   **c**

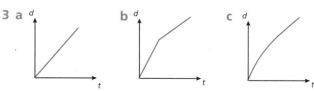

4 19 m

● Chapter 18

Now try this

1

x	0	1	2
y	−2	1	4

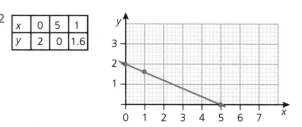

2

x	0	5	1
y	2	0	1.6

3

x	0	1	2
y	6	4	2

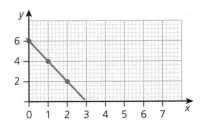

6 y-values: 18, 7, 0, −3, −2, 3, 12, 25

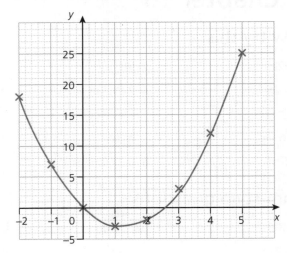

4 a

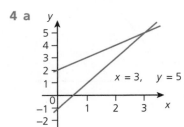

x = 3, y = 5

b

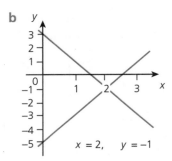

x = 2, y = −1

5 a y-values: 8, 4, 2, 2, 4, 8

b So that the minimum point on the graph can be plotted accurately, 1.75.

c

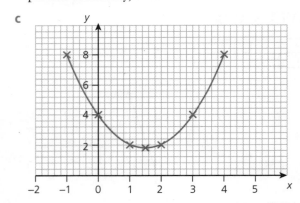

7 a x = 0 or 3

b x = −0.8 or 3.8 to 1 d.p.

c x = 0.6 or 3.4

8 a x = 0 or 2.5

b x = 2.3 or 0.2 to 1 d.p.

c x = −0.1 or 3.6

9 x = −0.6 or 1.6

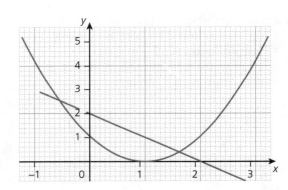

10

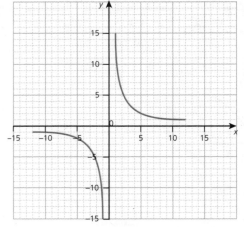

Cambridge IGCSE Mathematics Study and Revision Guide Second Edition © Brian Seager *et al.*, 2016

11 Roots $x = 0$ or ± 1.4 to 1 d.p; positive root 1.41 to 2 d.p.

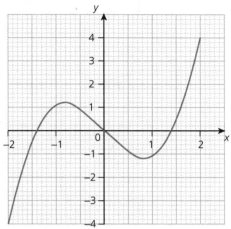

12 $x = 3.3$

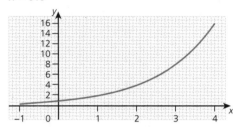

13

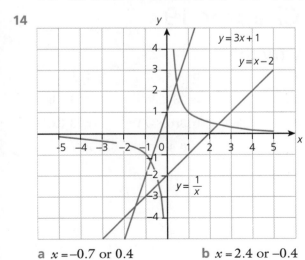

a $x = 0.3$ or 2.6 **b** $x = 0.6$ or 3.4 **c** $x = 0$ or 5

14

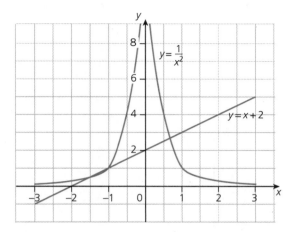

a $x = -0.7$ or 0.4 **b** $x = 2.4$ or -0.4

15 $x = -1.5$, -1 or 0.6

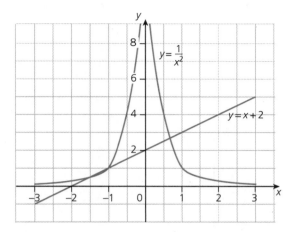

Answers read from graphs may not be exactly as given here but should approximate to these.

15 a 3 **b** 12

16 a -3 **b** -0.75 **c** -0.75

17 a -5 **b** 1

18 a 10 **b** -2 **c** 10

19 a $23\,\text{ms}^{-1}$ **b** $70\,\text{ms}^{-1}$

20 a $4\,\text{ms}^{-2}$ **b** $1.1\,\text{ms}^{-2}$

Exam-style questions

1 $x = 1\frac{1}{4}$, $y = 1\frac{3}{4}$

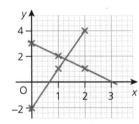

$x = 1\frac{1}{4}$, $y = 1\frac{3}{4}$

2 $x = -0.5$, $y = 2.5$

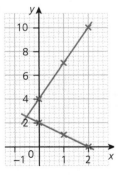

$x = -0.5$, $y = 2.5$

3 a

x	−1	0	1	2	3	4	5
y	−5	0	3	4	3	0	−5

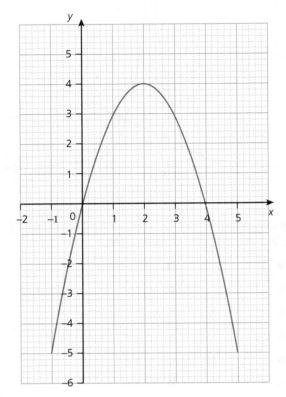

b (i) 2   **(ii)** 0.6 and 3.4

4 a T   **b** Q   **c** U

5 a (−2, −1) in table

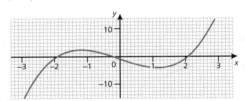

b −2 to −1.7, −0.4 to −0.2, 2.0 to 2.2

c $y = 2x + 2$ drawn, −2.3 to −2.0, −0.6 to −0.3, 2.5 to 2.8

6 $x = −1$, −0.6 or 1.6

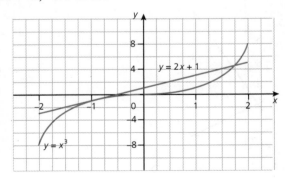

7 $x = −3.2$, −1.1 or 0.8

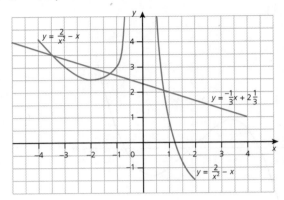

8 −1.6 ms⁻²

9 a

Time (*t* hours)	0	1	2	3	4	5
Number of bacteria (*n*)	20	60	180	540	1620	4860

b $n = 20 \times 3^t$

c

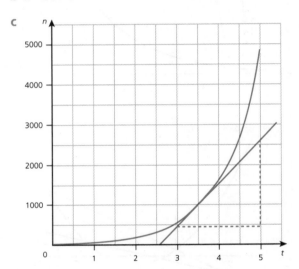

d (i) 1075

    (ii) The rate of increase in bacteria per hour

● Chapter 19

Now try this

1 a 36   **b** −14   **c** 1

2 a −21   **b** −3   **c** 3−8x

3 a 9   **b** 3(6x−7)   **c** 3(2x−9)

4 a 2   **b** 0.4   **c** 4

Cambridge IGCSE Mathematics Study and Revision Guide Second Edition © Brian Seager *et al.*, 2016

5 a 73

   b 1

   c $12x^2 - 2$

   d $3x^2 + 18x + 25$

   e $12x^2 + 3$

6 a 2   b –2   c –3   d 16

7 a 1   b –5

   c (i) $x - 2$ (ii) $x - 2$

8 a 50   b –18

   c (i) $10x + 10$ (ii) $10x - 13$

9 a 23   b 6

   c (i) $2x + 13$ (ii) $2x + 5$

10 a –23   b 22

   c (i) $-23 - 6x$ (ii) $10 - 6x$

11 a $4x^2 - 3$ b $16x^2 - 24x + 9$

12 a $2x^2 - 4x + 3$ b $4x^2$

13 a $\dfrac{x - 2}{7}$ b 0   c $\dfrac{-2}{7}$

14 a $4(x + 1)$ b 9   c –16

15 a $\dfrac{x}{3} - 2$ b $\dfrac{4x + 5}{2}$

   c $\dfrac{x}{12} + 2$ d $\sqrt{\dfrac{x - 3}{5}}$

Exam-style questions

1 a $\dfrac{x + 1}{2}$

   b $4x^2 - 4x$

2 a (i) 27   (ii) 3

   b (i) $2x^{\frac{2}{3}} - 5$ (ii) x^3

3 a –8   b $\dfrac{x + 5}{3}$ c $3x - 2$ d $x = 5$

4 a –1   b $\dfrac{3x - 5}{2}$

5 a $\dfrac{x - 1}{3}$ b 25

   c $18x^2 + 12x + 2$

   d $9x + 4$

6 a –1   b $\dfrac{x - 1}{2}$ c $3x + 15$

   d –9   e $\dfrac{2x}{3} - 9$

● Chapter 20

Now try this

1 a Acute   b Reflex   c Obtuse

   d Obtuse   e Reflex

2 a and d

3 Because, although the angles are equal, the sides are not proportional

● Chapter 21

Answers should be ±1 mm or ±1°.

1 a 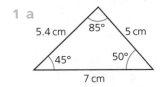   c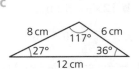

   b 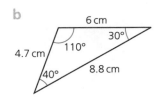   d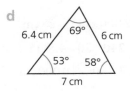

2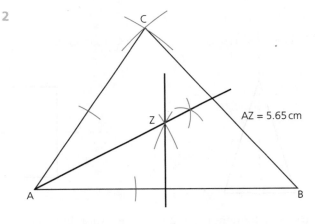

3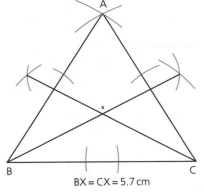

4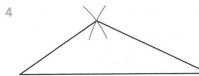

Cambridge IGCSE Mathematics Study and Revision Guide Second Edition © Brian Seager *et al.*, 2016

● Chapter 22

Now try this

1 $MN = 5 \times \frac{9}{6} = 7.5$

   $PQ = 4 \times \frac{6}{9} = \frac{8}{3} \ (= 2.7)$

2 a $65°$

   b $12 \times \frac{2}{3} = 8 \text{ cm}$

   c $7 \times \frac{3}{2} = 10.5 \text{ cm}$

3 $\frac{x}{x+7} = \frac{3}{8}$ $8x = 3x + 21$, $5x = 21$, $x = 4.2 \text{ cm}$

4 $76.5 \times \left(\frac{2}{3}\right)^2 = 76.5 \times \frac{4}{9} = 34 \text{ cm}^2$

5 $12\,800 \times \left(\frac{1}{8}\right)^3 = 25 \text{ cm}^3$

6 a $240 \div 50 = 4.8 \text{ cm}$

   b $4 \times 50^2 = 10\,000 \text{ cm}^2 = 1 \text{ m}^2$

7 Volume scale factor $= \frac{24}{81} = \frac{8}{27}$

   Length scale factor $= \sqrt[3]{\frac{8}{27}} = \frac{2}{3}$

   Area scale factor $= \left(\frac{2}{3}\right)^2 = \frac{4}{9}$

   $A = 540 \times \frac{4}{9} = 240 \text{ cm}^2$

● Chapter 23

Now try this

1 and 2   a Equilateral triangle, order 3

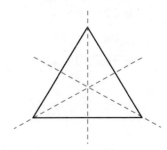

   b Isosceles triangle

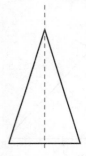

   c Rectangle, order 2

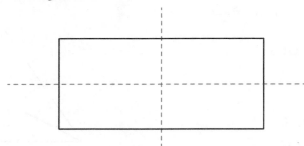

   d Parallelogram, order 2

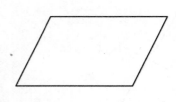

   e Rhombus, order 2

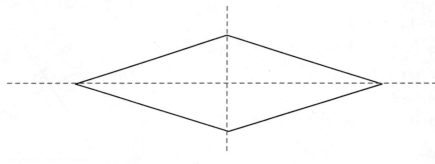

Shape d has no lines of symmetry, shape b has no rotational symmetry.

3 Equilateral triangular prism: a 4

   b 4, three of order 2 and one of order 3

   Square based pyramid: a 4   b 1 of order 4

4 Angle BAT $= 90 - 25 = 65° =$ angle ABT

   Angle ATB $= 180 - 2(65) = 50°$

● Chapter 24

Now try this

There are other correct reasons for the answers to questions 1 to 4.

1 $a = 104°$ (exterior angle = sum of opposite interior angles)

   $b = 52°$ (isosceles triangle and exterior angle = sum of opposite interior angles)

2 $c = 37°$ (alternate angles are equal)

   $d = 110°$ (exterior angle = sum of opposite interior angles)

3 $e = 17°$ (corresponding angles are equal)

   $f = 49°$ (alternate angles are equal)

4 $p + q = x$ (exterior angle = sum of opposite interior angles)

   $p = q$ (base angles of isosceles triangle are equal)

   So, $p = x°$

   $g = 180 - \frac{1}{2}x°$ (angles on a straight line = 180°)

5 Exterior angle = $360 \div 8 = 45°$. Interior angle = $180 - 45 = 135°$

6 $360 - (112 + 65 + 79) = 104°$

7 Exterior angle = $180 - 156 = 24°$. Number of sides = $360 \div 24 = 15$

8 $540 - (75 + 96 + 125 + 142) = 102°$

9 a $29°$ b $119°$

10 Angle CDA = $65°$ (opposite angles of a cyclic quadrilateral)

    Angle CDE = $180 - 65 = 115°$ (angles on a straight line)

11 Angle ADX = angle BCX (angles in the same segment)

    Angle DAX = angle CBX (angles in the same segment)

    Angle DXA = angle CXB (vertically opposite angles)

    (or any of these replaced by 'angle sum of triangle')

    Therefore, the triangles are similar (all angles are equal)

12 $a = 65°$, $b = 25°$, $c = 130°$

● Chapter 25

Now try this

1 Circle, centre P, radius 4 cm drawn

2 Perpendicular bisector of AB drawn and the side of it including A shaded

3 Bisector of the angle drawn

4 Shore line drawn running SE to NW. Locus of boat drawn – a line parallel to the shore and 500 m from it

● Chapter 26

Now try this

1 4 cm   2 1.5 m

3 1500 g   4 52 000 ml

5 153 mm   6 1.465 kg

7 1.34 litres   8 0.075 m

9 4000 mm²   10 15 000 cm²

11 1500 litres   12 5.254 m³

13 5 cm²   14 15 300 mm³

15 6 300 000 cm³   16 24 m³

● Chapter 27

Now try this

1 42 cm²   2 0.77 m²

3 251.3 m²   4 114.6°

5 15.3 cm   6 5 cm

7 265 cm³   8 2 cm

9 452.4 cm³   10 11.9 cm

11 184 cm³

12 a 4.8 cm   b 10 000 cm² = 1 m²

13 240 cm²

● Chapter 28

Now try this

1 a Gradient $= -1.5$, y-intercept $= 3$

  b Gradient $= 2$, y-intercept $= 2$

2 a $-\frac{1}{4}$ **b** $1\frac{1}{2}$

  c 2 **d** $-\frac{1}{2}$

3 a $(1, 0)$ **b** $(0, -2)$

  c $(0, -1)$ **d** $(0, 1)$

4 (i) a $(2, 3)$ **b** 5.7 to 1 d.p.

  (ii) a $\left(-1, 4\frac{1}{2}\right)$ **b** 6.7 to 1 d.p.

  (iii) a $\left(\frac{1}{2}, 3\right)$ **b** 13.6 to 1 d.p.

  (iv) a $\left(-3\frac{1}{2}, -1\frac{1}{2}\right)$ **b** 1.4 to 1 d.p.

  (v) a $\left(\frac{1}{2}, 0\right)$ **b** 6.7 to 1 d.p.

5 a $y = -1.5x + 3$ **b** $y = 2x + 2$

6 a $m = 4$, $c = -1$ **b** $m = 2$, $c = 3$

  c $m = 1$, $c = 0$ **d** $m = -1$, $c = 4$

  e $m = 1.5$, $c = 2$ **f** $m = \frac{1}{4}$, $c = -2.5$

7 a $y = 3x - 2$ **b** $y = -\frac{1}{2}x + 7$

  c $y = 2x + 8$ **d** $y = -4x - 11$

8 a $y = 3x - 3$ **b** $y = -2x + 5$

9 (a), (d) and (f) are parallel; (c) and (e) are parallel

10 $y = 2x + 7$ **11** $y = 3x - 9$

12 $y = -\frac{1}{2}x - \frac{1}{2}$ **13** $y = -3x + 14$

14 (b) and (c)   **15** $y = -\frac{1}{4}x + 3$

16 $y = 2x + 12$ **17** $y = -\frac{1}{3}x + 1$

Exam-style questions

1 a -2.5 **b** $y = -2.5x + 5$

2 a $m = -2$, y-intercept $= 4$

  b $y = -2x - 1$

3 $y = -2x + 6$

4 Parallel: $x + y = 2$

  Perpendicular: $x = y$

5 a $y = -x + 5$

  b $y = x - 7$

● Chapter 29

Now try this

1

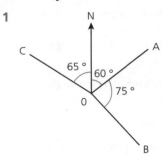

2 $307°$

3

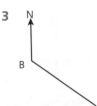

Exam-style questions

When you have to take a measurement from your diagram, answers within 2 mm or 2° of the answer given below are acceptable.

1 a

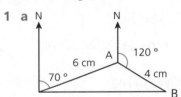

  b Distance $= 9.1$ km. Bearing $= 090°$

2 a

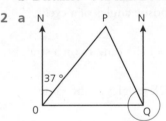

  b (i) PQ measures 3.4 cm on the diagram, so is $3.4 \times 2 = 6.8$ km

  (ii) Bearing is $341°$

3 a The bearing of Acacia from Corunna is $285°$

  b

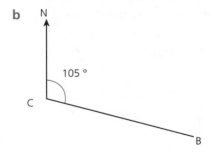

Cambridge IGCSE Mathematics Study and Revision Guide Second Edition © Brian Seager *et al.*, 2016

4 a (i) 232° **(ii)** 198°

b (i)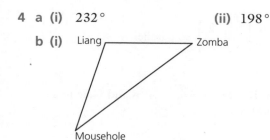

(ii) 14.2 km, any answer in the range 13.8 to 14.6 km is acceptable

● Chapter 30

Now try this

1 4.5 cm

2 3.6 cm

3 4.9 cm

4 a 3.87 cm **b** 29.0°

5 a 3.06 cm **b** 6.73 cm

6 a 3.77 cm **b** 46.7°

7 a 8.54 m **b** 1.07 m

8 64.6°

9 a 2.57 m **b** 66°

10 a (i) sin 80° **(ii)** sin 27°

(iii) sin 30° **(iv)** sin 120°

b (i) 13°, 167° **(ii)** 28°, 152°

(iii) 5°, 175° **(iv)** 56°, 124°

11 (i) −cos 80° **(ii)** −cos 27°

(iii) cos 30° **(iv)** cos 120°

Exam-style questions

1 a 4.69 cm **b** 59° and 31°

2 3.91 m

3 a 4.57 m **b** 66.8°

4 a 9.43 cm **b** 10.63 cm **c** 8.60 cm

5 59.4 m

6 a

y
1

$y = \sin x$

0

90   180 x

−1

$y = \cos x$

b 45°

● Chapter 31

Now try this

1 a 22.46 m **b** 30.4°

2 6.85 cm

3 331 m

Exam-style questions

1 a 17.7 m **b** 8.94 m

2 a 5.2 m **b** 3.1 m²

3 a 21.2 km **b** 155 km²

4 a 58° **b** 38.8 cm²

5 58°

● Chapter 32

Now try this

1 a (i) $\begin{pmatrix} 2 \\ 4 \end{pmatrix}$ **(ii)** $\begin{pmatrix} 3 \\ -3 \end{pmatrix}$ **(iii)** $\begin{pmatrix} 6 \\ -2 \end{pmatrix}$

(iv) $\begin{pmatrix} 2 \\ 5 \end{pmatrix}$ **(v)** $\begin{pmatrix} 5 \\ 1 \end{pmatrix}$ **(vi)** $\begin{pmatrix} 4 \\ 3 \end{pmatrix}$

b (i) $\begin{pmatrix} 7 \\ 6 \end{pmatrix}$ **(ii)** $\begin{pmatrix} -9 \\ -4 \end{pmatrix}$ **(iii)** $\begin{pmatrix} -6 \\ -8 \end{pmatrix}$

c (i) $\begin{pmatrix} 15 \\ 3 \end{pmatrix}$ **(ii)** $\begin{pmatrix} -8 \\ -6 \end{pmatrix}$ **(iii)** $\begin{pmatrix} -8 \\ -20 \end{pmatrix}$

2 a (i) $\begin{pmatrix} 4 \\ 4 \end{pmatrix}$ **(ii)** $\begin{pmatrix} 6 \\ 0 \end{pmatrix}$

b (i) $3a + 2b = \begin{pmatrix} 13 \\ 10 \end{pmatrix}$

(ii) $2a - 3b = \begin{pmatrix} 13 \\ -2 \end{pmatrix}$

3 a 5.4   **b** 5   **c** 8.5

4 a 7.6   **b** 7.8   **c** 3.6

5 a 2.2   **b** 12.0   **c** 9.4

    d 8.5   **e** 27.0

6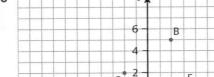

7 a $\overrightarrow{OA} = \begin{pmatrix} 7 \\ 4 \end{pmatrix}$, $\overrightarrow{OB} = \begin{pmatrix} 3 \\ 1 \end{pmatrix}$, $\overrightarrow{OC} = \begin{pmatrix} 5 \\ -2 \end{pmatrix}$,

$\overrightarrow{OD} = \begin{pmatrix} 4 \\ -4 \end{pmatrix}$, $\overrightarrow{OE} = \begin{pmatrix} -1 \\ -6 \end{pmatrix}$, $\overrightarrow{OF} = \begin{pmatrix} -2 \\ -4 \end{pmatrix}$,

$\overrightarrow{OG} = \begin{pmatrix} -6 \\ -3 \end{pmatrix}$, $\overrightarrow{OH} = \begin{pmatrix} -4 \\ 0 \end{pmatrix}$, $\overrightarrow{OI} = \begin{pmatrix} -7 \\ 3 \end{pmatrix}$,

$\overrightarrow{OJ} = \begin{pmatrix} -3 \\ 5 \end{pmatrix}$

b (i) $\begin{pmatrix} 11 \\ 4 \end{pmatrix}$ **(ii)** $\begin{pmatrix} 10 \\ -2 \end{pmatrix}$

(iii) $\begin{pmatrix} 11 \\ 1 \end{pmatrix}$ **(iv)** $\begin{pmatrix} 8 \\ -4 \end{pmatrix}$

(v) $\begin{pmatrix} -8 \\ -10 \end{pmatrix}$ **(vi)** $\begin{pmatrix} 1 \\ -9 \end{pmatrix}$

(vii) $\begin{pmatrix} -9 \\ -4 \end{pmatrix}$ **(viii)** $\begin{pmatrix} -9 \\ 2 \end{pmatrix}$

(ix) $\begin{pmatrix} -11 \\ 7 \end{pmatrix}$ **(x)** $\begin{pmatrix} 3 \\ 8 \end{pmatrix}$

8 $\overrightarrow{AB} = \begin{pmatrix} 6 \\ 9 \end{pmatrix} = 3\begin{pmatrix} 2 \\ 3 \end{pmatrix}$, so parallel

9 a $b - a$

    b $\overrightarrow{AC} = 4b - 4a = 4\,\overrightarrow{AB}$, so A, B and C lie in a straight line

Exam-style questions

1 a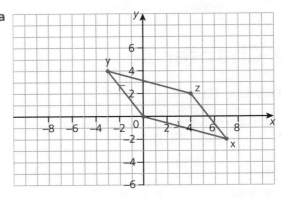

    b $YX = \begin{pmatrix} 10 \\ -6 \end{pmatrix}$

    c (i) $(4, 2)$ **(ii)** $ZO = \begin{pmatrix} -4 \\ -2 \end{pmatrix}$

2 $BC^2 = 3^2 + 5^2 = 34$

    $BC = \sqrt{34} = 5.83$ units to 2 d.p.

3 a $(4, 6)$ $(1, 3.5)$ $(0, -0.5)$

    b $\begin{pmatrix} -2 \\ -8 \end{pmatrix}$ $\begin{pmatrix} -1 \\ -4 \end{pmatrix}$

    c MN is parallel to BC and $BC = 2 \times MN$

4 a (i) $2b$ **(ii)** $a - b$ **(iii)** $b + \frac{1}{2}a$

    b $\frac{1}{3}(a + 2b)$

    c Collinear, and $OM = 1.5 \times OH$

5 a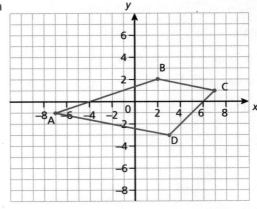

b (i) $\overrightarrow{AC} = \begin{pmatrix} 14 \\ 2 \end{pmatrix}$

(ii) $\overrightarrow{DB} = \begin{pmatrix} -1 \\ 5 \end{pmatrix}$

c $|\overrightarrow{AC}| = 14.1$

● Chapter 33

Now try this

1 a 2×4　　　**b** 4×2　　　**c** 3×4

2 $\begin{pmatrix} 5 & 8 & 10 \\ 11 & 6 & 9 \end{pmatrix}$

3 $\begin{pmatrix} 86 & 58 & 102 \\ 92 & 59 & 98 \\ 95 & 56 & 116 \end{pmatrix}$

4 a $\begin{pmatrix} 11 & 6 & 7 & 6 \\ 7 & 10 & 10 & 7 \end{pmatrix}$　　**b** $\begin{pmatrix} 37 & 12 \\ 6 & 35 \\ 25 & 36 \end{pmatrix}$

5 a $\begin{pmatrix} 2 & -2 \\ -1 & -2 \end{pmatrix}$　　**b** $\begin{pmatrix} -7 & 2 & 0 \\ -1 & 7 & -8 \\ -1 & 5 & -3 \\ 1 & -7 & 2 \end{pmatrix}$

6 a $\begin{pmatrix} 13 & 19 \\ 25 & 37 \end{pmatrix}$　　**b** $\begin{pmatrix} 21 & -2 \\ -7 & 18 \end{pmatrix}$

c Not possible　　**d** $\begin{pmatrix} 24 \\ 26 \end{pmatrix}$

e $(-11 \ -6)$　　**f** $(8 \ 5)$

g $\begin{pmatrix} 9 & 12 & 6 \\ 1 & -24 & 14 \end{pmatrix}$　　**h** $\begin{pmatrix} 10 & -7 & 16 \\ -5 & 28 & -15 \\ 10 & 24 & 32 \end{pmatrix}$

7 a 7　　**b** 4　　**c** -1　　**d** 3
　　e 26　　**f** -7　　**g** 23　　**h** -2

8 a $\begin{pmatrix} 3 & -5 \\ -4 & 7 \end{pmatrix}$　　**b** $\begin{pmatrix} 5 & -3 \\ -3 & 2 \end{pmatrix}$

c $\begin{pmatrix} 4 & -9 \\ -3 & 7 \end{pmatrix}$　　**d** $\begin{pmatrix} -2 & -3 \\ -5 & -8 \end{pmatrix}$

e $\begin{pmatrix} 1.25 & -0.5 \\ -2 & 1 \end{pmatrix}$　　**f** $\begin{pmatrix} 0.5 & -0.4 \\ -0.5 & 0.6 \end{pmatrix}$

g $\begin{pmatrix} -3 & -3.5 \\ -2 & -2.5 \end{pmatrix}$　　**h** $\begin{pmatrix} -2.5 & 4 \\ 2 & -3 \end{pmatrix}$

Exam-style questions

1 a $AB = \begin{pmatrix} 19 \\ 14 \end{pmatrix}$　　**b** $A^{-1} = \begin{pmatrix} 1 & -1 \\ -\frac{1}{2} & \frac{3}{4} \end{pmatrix}$

2 a $M = \begin{pmatrix} -3 & 2 \\ -1 & -2 \end{pmatrix}$　　**b** $N = \begin{pmatrix} \frac{1}{4} & \frac{1}{4} \\ -\frac{1}{8} & \frac{3}{8} \end{pmatrix}$

3 a $x = -4$

b $|R|$, the determinant of R, is zero

c $Q^{-1} = \begin{pmatrix} \frac{3}{5} & \frac{2}{5} \\ \frac{2}{5} & \frac{3}{5} \end{pmatrix}$

4 a $p = -3, q = -17$　　**b** $X^{-1} = \begin{pmatrix} \frac{1}{5} & \frac{2}{5} \\ \frac{4}{5} & \frac{3}{5} \end{pmatrix}$

5 a The number of columns in A is not the same as the number of rows in B.

b (i) $B^2 = \begin{pmatrix} 17 & -32 \\ -16 & 33 \end{pmatrix}$　　**(ii)** $B^{-1} = \begin{pmatrix} \frac{5}{7} & \frac{4}{7} \\ \frac{2}{7} & \frac{3}{7} \end{pmatrix}$

● Chapter 34

Now try this

1

2

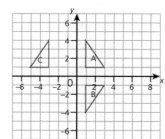

3 a Reflection in y-axis

b Reflection in $x = -2$

c Enlargement with centre $(0, 3)$, scale factor 3

d Enlargement with centre $(0, 3)$, scale factor $\frac{1}{3}$

e Translation by $\begin{pmatrix} -5 \\ 2 \end{pmatrix}$

f Translation by $\begin{pmatrix} 2 \\ -6 \end{pmatrix}$

g Translation by $\begin{pmatrix} 4 \\ 0 \end{pmatrix}$

h Rotation through 90° clockwise about $(1.5, -1.5)$

i Translation by $\begin{pmatrix} -4 \\ 0 \end{pmatrix}$

j Reflection in $y = x$

k Enlargement with centre $(-1, 6)$, scale factor 3

4

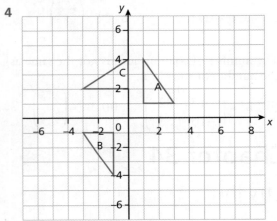

5 $(4, 0)$; 3

6

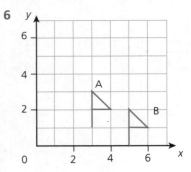

7 Translation of $\begin{pmatrix} 3 \\ 1 \end{pmatrix}$

8 Rotation through 180° about $(3, 2)$

9 Reflection in $y = x$

10 Reflection in $y = -x$

11 a A′ $(2, 0)$, B′$(4, 0)$, C′$(4, -2)$, D′$(2, -2)$

b A″$(0, -2)$, B″$(0, -4)$, C″$(2, -4)$, D″$(2, -2)$

c A‴$(0, -2)$, B‴$(0, -4)$, C‴$(-2, -4)$, D‴$(-2, -2)$

d $R = \begin{pmatrix} 0 & -1 \\ -1 & 0 \end{pmatrix}$, $S = \begin{pmatrix} 0 & 1 \\ 1 & 0 \end{pmatrix}$

e $(0, -2)$, $(0, -4)$, $(-2, -4)$ $(-2, -2)$, reflection in the line $y = -x$

f $(0, 2)$, $(0, 4)$, $(2, 4)$, $(2, 2)$, reflection in the line $y = x$

12 a (i) $(-2, -1)$, $(-4, -1)$, $(-4, -4)$

(ii) Rotation of 180° about the origin

b (i) $(2, -1)$, $(4, -1)$, $(4, -4)$

(ii) Reflection in the x-axis

Exam-style questions

1 a Rotation through 90° clockwise about $(0, 0)$.

b and c

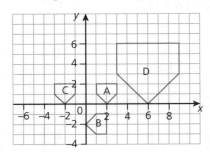

2 Enlargement with centre $(0, 0)$ and scale factor -1

3 a $\begin{pmatrix} -1 & 0 \\ 0 & 1 \end{pmatrix}$

b $\begin{pmatrix} 0 & 1 \\ -1 & 0 \end{pmatrix}$

c Reflection in the line $y = x$

d QP gives a reflection in the line $y = x$

PQ gives a reflection in the line $y = -x$

The inverse of a reflection in the line $y = x$ is itself a reflection in the line $y = x$, not a reflection in the line $y = -x$

4 a Reflection in the line $x + y = 0$, i.e. $y = -x$

b $\begin{pmatrix} 0 & 1 \\ -1 & 0 \end{pmatrix}$

c (i) Scale factor is -2

(ii) Centre is $(1, 5)$

d (i) Coordinates are $(8, 6)$, $(10, 6)$, $(11, 15)$

(ii) 9 sq units

(iii) $\begin{pmatrix} \frac{1}{2} & \frac{-1}{6} \\ 0 & \frac{1}{3} \end{pmatrix}$

Cambridge IGCSE Mathematics Study and Revision Guide Second Edition © Brian Seager *et al.*, 2016

● Chapter 35

Now try this

1 a $1-(0.05+0.2+0.25+0.35)=0.15$

b $1-0.25=0.75$

c Orange $0.05\times20=1$

White $0.15\times20=3$

Yellow $0.2\times20=4$

Green $0.25\times20=5$

Red $0.35\times20=7$

2 a

Top	Middle	Bottom
A	B	C
A	C	B
B	A	C
B	C	A
C	A	B
C	B	A

b $\frac{2}{6}=\frac{1}{3}$

3 0.148

Yes, the results do suggest the spinner is biased. This a large number of trials and there should be roughly the same frequency for each number, but there are nearly twice as many 4s as 3s

4 a 43 **b** 0.288, to 2 d.p. or more

● Chapter 36

1

Main course	Dessert
CFR	BC
CFR	IC
CFR	BF
PCM	BC
PCM	IC
PCM	BF

2

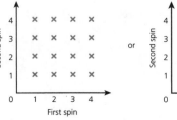

a $\frac{1}{16}$ **b** $\frac{3}{16}$

3 a

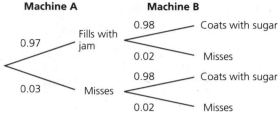

b 0.0006 **c** 0.0488

4 a $\frac{8}{125}$ **b** $\frac{27}{125}$ **c** $\frac{54}{125}$

5 a $\frac{1}{15}$ or equivalent

b $\frac{31}{45}$ or equivalent

6 a $\frac{4}{25}$ **b** $\frac{23}{50}$

7 $\frac{8}{15}$

8 $\frac{10}{21}$

Exam-style questions

1 a 0.3 added to first set of branches, 0.8 and 0.2 to each of second sets.

b (i) 0.56

(ii) 0.38

2 a $\frac{6}{10}$ and $\frac{4}{10}$ on first set of branches, $\frac{2}{3},\frac{1}{3},\frac{1}{4},\frac{3}{4}$ on second set with 'goes', 'not go' or other suitable labels.

b (i) $\frac{2}{5}$ or equivalent

(ii) $\frac{3}{10}$ or equivalent

3 a $\frac{1}{7}$ or equivalent

b $\frac{31}{105}$

c $1-$ (b) $=\frac{74}{105}$

4 a $\frac{1}{36}$

b (i) $\left(\frac{5}{6}\right)^{n}$ **(ii)** $1-\left(\frac{5}{6}\right)^{n}$

5 a Box A   Box B

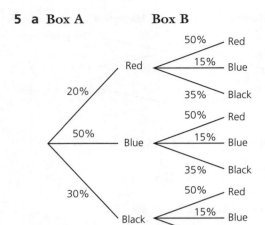

b 10%   **c** 28%

● Chapter 37

Now try this

1 a $190 - 171 = 19$ **b** 171 cm

   c 174 cm   **d** Mean $= \dfrac{1980}{11} = 180$ cm

2

Weight (g)	Frequency	
454	7	3178
455	10	4550
456	25	11 400
457	30	13 719
458	8	3664
	80	36 511

  a 457 g   **b** 456 g

  c Mean $= \dfrac{36,511}{80} = 456.3875$ g

  d $458 - 454 = 4$

3

Time (t, min)	Frequency (f)	Mid-value	Mid-value × frequency
$0 \leq t < 6$	0	3	0
$6 \leq t < 12$	4	9	36
$12 \leq t < 18$	13	15	195
$18 \leq t < 24$	16	21	336
$24 \leq t < 30$	6	27	162
$30 \leq t < 36$	1	33	33
Total	40		762

Mean $= \dfrac{762}{40} = 19.05$ min

4

Height (x, m)	Number of students (f)	Mid-interval value	Mid-interval value × frequency
$2.0 < x \leq 2.2$	1	2.1	2.1
$2.2 < x \leq 2.4$	14	2.3	32.2
$2.4 < x \leq 2.6$	31	2.5	77.5
$2.6 < x \leq 2.8$	29	2.7	78.3
$2.8 < x \leq 3.0$	25	2.9	72.5
$3.0 < x \leq 3.2$	17	3.1	52.7
$3.2 < x \leq 3.4$	3	3.3	9.9
Total	120		325.2

Mean $= \dfrac{325.2}{120} = 2.71$ m

5

Weight (x, g)	Number of apples (f)	Mid-interval value	Mid-interval value × frequency
$50 < x \leq 60$	23	55	1265
$60 < x \leq 70$	34	65	2210
$70 < x \leq 80$	58	75	4350
$80 < x \leq 90$	20	85	1700
$90 < x \leq 100$	15	95	1425
Total	150		10 950

Mean $= \dfrac{10,876}{150} = 73$ g

Exam-style questions

1 Mean $= \dfrac{19}{20} = 0.95$ spelling mistakes per page

2

Number of screws (x)	Number of boxes (f)	Number of screws × number of boxes
76	5	380
77	9	693
78	16	1248
79	22	1738
80	32	2560
81	12	972
82	3	246
83	1	83
Total	100	7920

Mean $= \dfrac{7920}{100} = 79.2$ screws

Cambridge IGCSE Mathematics Study and Revision Guide Second Edition © Brian Seager *et al.*, 2016

3

Weight (x, kg)	Frequency (f)	Mid-value	Mid-value × frequency
0 < x ≤ 2	3	1	3
2 < x ≤ 4	9	3	27
4 < x ≤ 6	16	5	80
6 < x ≤ 8	8	7	56
8 < x ≤ 10	3	9	27
10 < x ≤ 12	1	11	11
12 < x ≤ 14	0	13	0
Total	40		204

$$\text{Mean} = \frac{204}{40} = 5.1\,\text{kg}$$

4

Time (x, min)	Frequency (f)	Mid-value	Mid-value × frequency
0 < x ≤ 1	4	0.5	2.0
1 < x ≤ 2	20	1.5	30.0
2 < x ≤ 3	12	2.5	30.0
3 < x ≤ 4	8	3.5	28.0
4 < x ≤ 5	17	4.5	76.5
5 < x ≤ 6	24	5.5	132.0
6 < x ≤ 7	36	6.5	234.0
7 < x ≤ 8	19	7.5	142.5
8 < x ≤ 9	8	8.5	68.0
9 < x ≤ 10	2	9.5	19.0
Total	150		762.0

$$\text{Mean} = \frac{762}{150} = 5.08\,\text{min}$$

● Chapter 38

Now try this

1 a

Score	Tally	Frequency
35	I	1
36	I	1
37	JHT	5
38	JHT IIII	9
39	JHT IIII	9
40	JHT	5

b

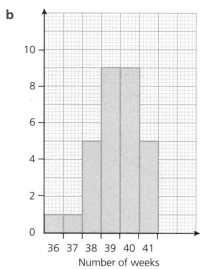

Number of weeks

2

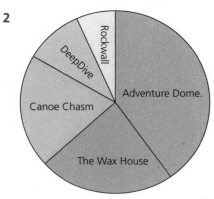

3 a 11   **b** 3

c

Monday	⬭ ⌒
Tuesday	⬭ ⬭ ⬭
Wednesday	⬭ ⬭ ⌒
Thursday	⬭ ⬭
Friday	⬭ ⬭ ⬭
Saturday	⬭ ⬭ ⬭ ⬭ ⬭ ⌒

4 Method 2 as the sample would be randomly selected

Method 1 fails because it is chosen on Saturday, which may not be representative of other days

Method 3 fails because the area may not be representative of the town

5 Any fair unbiased question acceptable. Examples include:

Do you watch television?

What is your favourite television programme?

How many hours per week do you watch television: 0–5, 6–10, 11–15, 16–20, 21–25, more than 25?

6 a and b

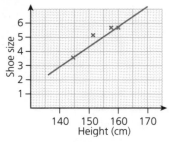

Fairly strong positive correlation

c 4

7 a and b

Fairly strong negative correlation

c 18 000

8 a 42, 66, 55, 15, 6, 1.6

b

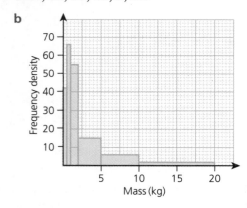

9 a 1.6, 4, 5, 2.5, 0.85

b

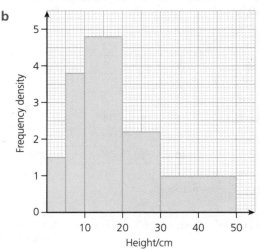

c 18.5 cm

Exam-style questions

1 a

Number of different vowels	Frequency
1	8
2	13
3	5
4	1
5	1

b

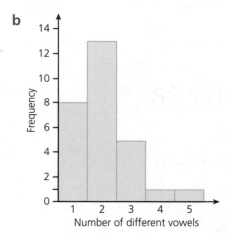

2

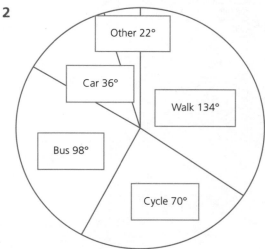

Cambridge IGCSE Mathematics Study and Revision Guide Second Edition © Brian Seager *et al.*, 2016

3 a $450

b

Housing	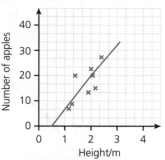
Fuel	
Food	
Clothes	
Travel	
Other	

c $200

4 a Possibly embarrassing, could be changed to give categories to select.

b Assumes respondent went on holiday. Could be changed to 'Did you go on holiday last year?' and then ask 'Where?'

c Assumes respondent reads newspapers. Could be changed to 'Did you read any newspapers this week?' and then ask 'How many?' or 'Which ones?'

d Do you enjoy sport?

e Possibly embarrassing, could be changed to give categories to select.

f How many children do you have? Assumes respondent has children. Could be changed to 'Do you have any children?' and then ask 'How many?'

5 Any valid reason. Examples include: 100 may or may not be a suitable sample size depending upon the population of Ayton. Some people who use taxis may not have a telephone and would therefore be excluded. People may use Ayton Taxis usually but may not have had to do so in the past week.

6 a (i) and (iii)

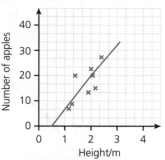

(ii) Positive correlation

(iv) 4 m is too far outside the range of the data

b Mid-interval values: 55, 65, 75, 85, 95

$w \times f$: 1265, 2730, 3750, 1700, 1425, 10 870

Mean: 72.5 g to 3 s.f.

7 a

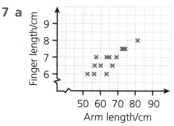

b Positive correlation – as arm length increases, so does finger length

c Between 6.8 cm and 7 cm

8 Frequency densities: 2.8, 4.1, 5.9, 3.5, 0.53

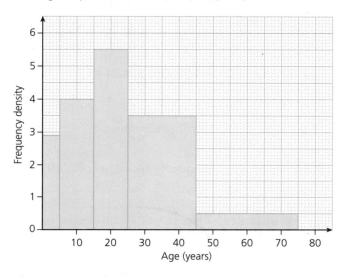

9 Frequency densities: 12, 16, 9.5, 4, 1.5

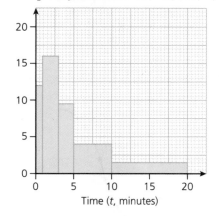

10

Time (minutes)	0–20	20–30	30–45	45–60	60–90
Frequency	28	60	132	108	60

● Chapter 39

Now try this

1 a Cumulative frequencies: 8, 23, 56, 78, 95, 100

b

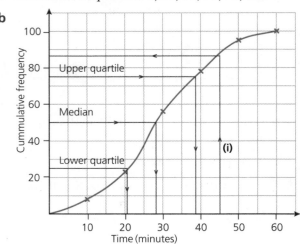

c (i) 13 or 14, **(ii)** 28, **(iii)** 38 − 21 = 17

2 a 84

b Median = 56, inter-quartile range
= 70 − 41 = 29

3 Cumulative frequencies: 4, 16, 32, 41, 47, 49, 50

a

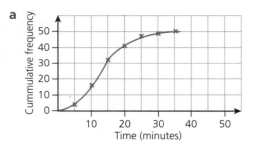

b Median = 13 minutes

4 a Cumulative frequencies: 5, 12, 28, 52, 71, 80

b

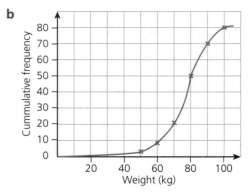

c Inter-quartile range = 82 − 68 = 14

d 45 sheep sent to market

Exam-style questions

1 a (i) [9], 22, 43, 60, 70, 78, 80

(ii)

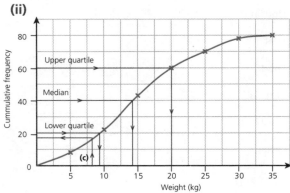

b (i) 14 to 14.8 kg

(ii) 10.5 to 11 is acceptable

c 63 to 65 is acceptable

2 a (i) [0], [5], 30, 75, 157, 190, [200]

(ii)

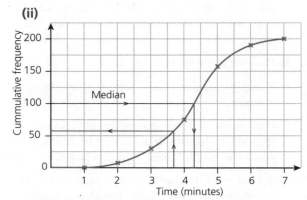

b (i) 4.2 to 4.5 minutes

(ii) 0.69 to 0.75

3 a

Time (t, min)	$t \leq 20$	$t \leq 40$	$t \leq 60$	$t \leq 80$	$t \leq 100$
Cumulative frequency	4	23	53	71	80

b

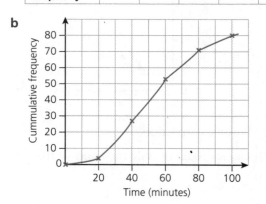

c (i) Median = 49 minutes

(ii) Inter-quartile range = 68 − 33 = 35

Cambridge IGCSE Mathematics Study and Revision Guide Second Edition © Brian Seager *et al.*, 2016

Index

Cambridge IGCSE Mathematics Study and Revision Guide Second Edition © Brian Seager *et al.*, 2016